The Belt Conveyor

The Belt Conveyor

A Concise Basic Course

D.V. Subba Rao

Formerly Head of the Department of
Mineral Beneficiation and Materials Handling
S D S Autonomous College, Garividi
Vizianagaram District, Andhra Pradesh, India

CRC Press
Taylor & Francis Group
Boca Raton London New York

CRC Press is an imprint of the
Taylor & Francis Group, an **informa** business

CRC Press/Balkema is an imprint of the Taylor & Francis Group, an informa
business

© 2021 Taylor & Francis Group, London, UK

Typeset by codeMantra

Library of Congress Cataloging-in-Publication Data
Applied for

Published by: CRC Press/Balkema
Schipholweg 107C, 2316 XC Leiden, The Netherlands
e-mail: Pub.NL@taylorandfrancis.com
www.routledge.com – www.taylorandfrancis.com

ISBN: 978-0-367-53570-4 (hbk)
ISBN: 978-0-367-54441-6 (pbk)
ISBN: 978-1-003-08931-5 (eBook)

DOI: 10.1201/9781003089315
https://doi.org/10.1201/9781003089315

Dedicated to my loving daughters

Radha   Lalitha

Contents

List of Figures

List of Tables

Preface

This book *The Belt Conveyor: A Concise Basic Course* is written with an intention to make the reader to have an overall understanding of the belt conveyor without going through the details of technology, design and calculations concerning belt conveyor. It is a fact that so much of literature is available on belt conveyors. Literature including volumes of research papers has been developed for specific purposes. To the best of my knowledge, there is no book on all basics dealt in brief as in this book.

This book is a concise basic book on belt conveyor containing a description of all parts of a belt conveyor, their functions and different types presented one after the other in a simplest possible way and easy-to-understand manner for a beginner or a new reader. The book is planned in such a way that even non-engineering persons will be able to go through the entire book in 3 to 4 hours and have enough understanding about belt conveyors.

In this book, properties of bulk materials are explained elaborately as the properties play an important role in conveyor performance. The properties are not only important in belt conveyor, but also in any bulk material handling equipment. A short chapter on belt tension has been included in this book to give an idea to the reader about various tensions that come to existence while belt conveyor is in operation, which are to be taken into account while designing and analysing the belt conveyor. This will indicate the need of referring detailed literature. Of course, this book contains no design aspects.

In the last chapter 'Troubleshooting', the troubles, probable causes for such troubles and corrective measures are given, through which interrelationships of different components can be understood. The purpose of this chapter is to provide an idea about each component for the overall performance of the belt conveyor during its operation.

I hope this book will be best appreciated by the beginners, those who do not know anything about belt conveyor and want to know, and those who enter into a bulk material handling plant where they are supposed to look after belt conveyors without having studied earlier.

I sincerely appeal to the readers to communicate their feedback preferably by email on the contents and suggestions for improvement, which will be useful to incorporate in future editions.

D. V. Subba Rao
Formerly Head
Department of Mineral Beneficiation
and Materials Handling
S D S Autonomous College, Garividi
Vizianagaram District, Andhra Pradesh, India
dvsubbarao3@rediffmail.com
dvsubbarao3@gmail.com
Mobile : +91 81791 93942

Acknowledgements

I am grateful to Sri T. S. Kasturi, former general manager, Neyveli Lignite Corporation, Tamil Nadu, India, an expert in belt conveyor and consultant for many bulk material industries in India. I learned the subject of belt conveyor technology from him. He took me to several belt conveyors at Visakhapatnam Steel Plant and Ferro Alloys Corporation, Garividi, and explained several aspects by making me practically observe all those conveyors. Under his able guidance, I prepared the reading material for the semester paper "Belt Conveyor Technology" to our B.Sc. Materials Handling students. Much of the material has been supplied by him. I taught the paper for 20 years for our students. Excellent feedback received from them and their colleagues encouraged me to write this book.

I am thankful to my college management, who gave me the opportunity of heading the Department of Materials Handling, which made it compulsory for me to learn the belt conveyor technology. The help extended by my colleagues Sri Y. Ramachandra Rao, Sri K. Satyanarayana and Sri K. Ganga Raju is always remembered. I am thankful to my colleagues at SDS Autonomous College for their continuous support.

I thank Dr. T. C. Rao, former director, Regional Research Laboratory, Bhopal, and Professor and Head of the Department of Fuel and Mineral Engineering, Indian School of Mines, Dhanbad; Sri A. L. Mohan, former principal, SDS Autonomous College, Garividi; and Dr. C. Raghu Kumar, Head, Ore Beneficiation Technology, Tata Steel Ltd., Jamshedpur, for their continuous encouragement and support.

I am indebted to my students who raised questions in the class and expressed their difficulty to understand when I explain certain aspects. This made me to think of writing this book presenting the basics in an easy-to-understand way.

Discussions with my following students working in mineral industries in India and abroad helped me to write this book:

1. Dr. K. Srinivas, Section Head, Maaden Phosphate Company, Saudi Arabia
2. Dr. Y. Ramamurthy, Principal Researcher, Tata Steel Ltd., Jamshedpur
3. Mr. D. P. Chakravarthy, Sr. Manager, Rashi Steel & Power Ltd., Bilaspur
4. Mr. A. Jagga Rao, Asst. Manager, Trimex Sands Pvt. Ltd., Srikakulam, A.P
5. Dr. Sunil Kumar Tripathy, Principal Researcher, Tata Steel Ltd., Jamshedpur
6. Mr. G. Satish Kumar, Dy. Manager, Weir Minerals India Pvt. Ltd., Visakhapatnam
7. Mr. Suryanarayan Bisoyi, Engineer-Minerals, Indian Rare Earths, Chavara, Kerala

8. Mr. R. Satyanarayana, Sr. Service Engineer, Metso India Pvt. Ltd., Visakhapatnam
9. Mr. M. Varma Raju, Jr. Manager, R&D, JSW Steels Ltd., Vidyanagar, Karnataka
10. Mr. T. Trinadha Rao, Mineral Process Engineer, Usha Martin Ltd., Jamshedpur

Special thanks are due to my friend Mr. P. V. Narasimha Rao, former Procurement Officer, APCO, for editing the script.

Without the understanding and support of my wife Krishna Veni, this book would not have been possible to write.

Matter relating to properties of bulk materials has been taken from the book *Mineral Beneficiation: A Concise Basic Course* written by me and *Bulk Solids Handling: An Introduction to the Practice and Technology* edited by C. R. Woodcock and J. S. Mason published in the USA by Chapman and Hall, 1987.

The reading material prepared by me for the "belt conveyor technology" paper for our students in 1993 with the help of Sri T. S. Kasturi, and the following two books have been taken as primary sources for planning and writing this book.

1. *Conveyor Components, Operation, Maintenance, Failure Analysis,* written by T.S.Kasturi, M.C.Ranga Rajan & T.Krishna Rajan and published by PR Business Services, No.49, K.B.Dhasan Road, Alwarpet, Madras, June 1994.
2. *Belt Conveyors for Bulk Materials,* Prepared by the Engineering Conference of the Conveyor Equipment Manufacturers Association, 7 th Edition, 15 April 2014.

Books referred have been listed in bibliography.

It has been a pleasure to work again with the Taylor & Francis group in publishing this book.

I gratefully acknowledge the following organizations for permitting me to use respective equations/tables/figures/photographs:

Conveyor Equipment Manufacturers Association (CEMA)	Horsepower equation
	Effective tension equation
	Wrap factor and frequency factor formulae
	Table of flowability
	Spiral and bin lowering chutes
	Rock ladder and telescopic chute
	Drive equipment assemblies
Fenner Dunlop Engineered	Belt weave types
Conveyor Solutions	Solid-woven belt
Flexco Steel Lacing Co.	Mechanical fasteners
MELCO Conveyor Equipment (Pvt) Ltd	Conveyor structure
Cleveland Oak	Multiple-axis training roller
Rexnord Corporation	Belt conveyor with structure
ABB Motors and Mechanical Inc.	Lagged pulley
	Spiral drum pulley
	Wing pulleys
	Types of grooves on lagging

Chapter 1

Belt conveyor components

A belt conveyor is a simple piece of equipment and the most common means of transportation of bulk materials and is capable of carrying a greater diversity of products at the rates of thousands of tons per hour in a continuous and uniform stream over long distances than any other kind of continuously operating mechanical conveyors. A typical belt conveyor consists of two drum pulleys fitted with a continuous loop of an endless belt, known as the conveyor belt, running around two pulleys and resting on troughing idlers for the carrying run and on return idlers for the return run. One of the pulleys is powered, driven by an electrical motor, which moves the belt and the material on the belt forward. The powered pulley is called the drive pulley, while the un-powered pulley at the dead end is called the tail pulley. Today's conveyors consist of dozens of other components, each one specially designed for moving different materials.

The basic design of a belt conveyor is so robust that it will convey material under the most adverse conditions such as overloaded, flooded with water, buried in fugitive material, or abused in any number of other ways.

Any stray material that escapes from a conveyor belt at a place other than its normal discharge point is called fugitive material. It arises as spillage and leakage from transfer points, carryback or airborne dust that has been carried off the belt by air currents and the impact forces of feeding. The material, which adhered to the belt and other conveyor components, cleaned by the cleaners and sent back along with the discharge material is called carryback.

The advantages of belt conveyors are their superiority, reliability, versatility, economy and safety of operation, and practically unlimited range of capacities. Belt conveyors can receive material from one or more locations and discharge material to points anywhere along its length. Belt conveyors suit well in tunnels beneath stockpiles to reclaim and blend various materials from different stockpiles. Belt conveyors with stackers and reclaimers are found as the only practical means for large-scale operations.

A stacker is a boom-mounted belt conveyor that discharges the material at the end of the boom which can drop material onto a stockpile for storage. A rotating bucket wheel can also be equipped at the discharge end so that the belt conveyor discharges the material into the buckets of bucket wheel, and as it rotates, it will discharge or drop the material onto a stockpile. The same stacker with bucket wheel rotating in reverse direction reclaims the material from the stockpile and feeds it to the belt conveyor for transportation to another destination.

Environmental acceptability is an added advantage for the selection of belt conveyors over other means of transportation. Belt conveyors with auxiliary equipment can also perform numerous process functions during various stages of processing. Sorting and hand-picking of the material over moving belt is the best example.

Belt conveyors are suitable for transportation of material in open-cast/underground mines, steel plants, beneficiation plants, port trusts, thermal power plants, cement plants, chemical plants, fertilizer plants, etc. Thus, wherever continuous material transportation is involved, belt conveyors form the main artery of the transportation system. Belt conveyors have the adaptability to work in different types of terrain and different gradients and have flexible path of travel, and the length of the routes can be extended repeatedly as per the requirement. In the case of mining, where due to the advancement of mine, the transport system has to be relocated frequently, the conveyor system lends itself easily. In a mine, belt conveyors are used for the transportation of the mined material to a beneficiation plant, stockpiling and reclaiming of concentrated product, disposal of waste, transport of product to the point of use in a process plant or loading onto a ship for sale to an overseas user.

The lowest transportation and maintenance costs, low labour and low energy requirements are the exceptional features with belt conveyors as compared to the most other means of transportation of bulk materials. The dramatic increase in the operating costs of other means of transport has placed conveyors in an extremely favourable position for applications that were not considered previously.

Materials that range from very fine, dusty chemicals to large, lumpy ore and coal are transported by belt conveyors. Closely sized or friable materials are carried with minimum degradation. Materials causing sticking or packing if transported by other means are often handled successfully on belt conveyors. Hot materials such as coke, sinter, iron ore pellets, direct reduced iron and hot briquetted iron are conveyed successfully.

Automatic sampling devices are arranged in the conveying system from where accurate samples of the material can be obtained by cutting through the stream of material flowing from one conveyor to the next or to the next operation. Materials can also be weighed accurately and continuously while the material is transported on the conveyor belt. It is one of the highlights of a belt conveyor.

Innumerable uses and applications of the belt conveyor make this device useful and reliable to carry out several major operations in industries. Today the use of belt conveyors is an important and broadly accepted means of long-distance transportation of bulk materials. They are also used economically in steel plants for transporting materials between two processing units at a wide range of rates, sometimes even at a very low rate. Belt conveyors provide a continuous flow of material while avoiding the confusion, delays and safety hazards of rail and road traffic in steel plants. Belt conveyors can be operated continuously, round the clock and 24×7 hours when needed without loss of time for loading and unloading or empty return trips.

The demand for ever-increasing capacities and ever-longer conveying lengths has accelerated the development of the belt conveyor systems having far greater complexity employing higher degrees of automated control. New materials are being developed, and new conveying systems are being planned and tested especially for meeting the challenges with regard to the environment control.

1.1 Components of a belt conveyor

The essential components of a typical belt conveyor are as follows:

1. The **belt**, which forms the moving and supporting surface on which the conveyed material is placed. The belt not only carries the material, but also transmits the pull. It is the tractive element.
2. The **idlers/rollers**, which form the support for the troughed carrying strand of the belt and the flat return strand.
3. The **pulleys**, which support and direct the belt and control its tensions.
4. The **drive**, which imparts power through one or more pulleys to move the belt and its load. Usually, the drive is an electric motor with reduction gears.
5. The **structure**, which supports and maintains the alignment of idlers, pulleys and drive.

In addition to these five components, necessary ancillary equipment is also installed in every conveyor to improve the system's operation. The belt conveyor should be properly designed to the requirements on how the conveyor is best fed and discharged, accessibility for operation and maintenance, electrical starting and stopping needs, and many other factors must be studied and carefully coordinated. The following are the components of a belt conveyor:

1. Top strand of the belt
2. Bottom strand of the belt
3. Carrying troughing idlers/rollers
4. Transition idlers/rollers
5. Impact idlers/rollers
6. Carrying training idlers/rollers
7. Return idlers/rollers
8. Return training idlers/rollers
9. Drive or head pulley
10. Tail pulleys
11. Snub pulleys
12. Bend pulleys
13. Take-up pulleys
14. Drive motor
15. Counterweights
16. Vertical take-up arrangement
17. Feed chute
18. Skirt plates for guiding the material
19. Belt cleaners for the carrying side of the belt
20. Belt cleaner (plough type) for the bottom side of the belt
21. Pulley cleaner
22. Discharge chute
23. Dribble chute.

The components of the belt conveyor are shown in Figure 1.1 along with the notation for the different lengths as given below.

L_P – Pulley centre distance
L_C – Length of the conveyor
L_{RT} – Idler/roller spacing top strand
L_{RB} – Idler/roller spacing bottom strand
h – Height of drop of material

In a simple system, as shown in Figure 1.1, an endless belt, made up of carcass impregnated with rubber and covered by rubber on both sides and edges, wraps around two pulleys, a drive pulley (9) and a tail pulley (10). The top strand of the belt (1) carries the material and is usually troughed and transforms to flat shape as it reaches the drive pulley. The design of the trough depends on the quantity of the load to be carried. The bottom strand of the belt (2) is a return strand and assumes flat shape after the drive pulley, travels round the tail pulley and then slowly takes the trough shape on top run and continues to be in trough shape till it reaches the drive pulley.

The top strand of the belt is supported by a set of idlers/rollers called carrying troughing idlers/rollers (3). By the time the belt reaches the troughing idlers/rollers, the belt takes full trough shape. Between the tail pulley and the load zone, idlers/rollers with increasing trough angles are fixed in order to transform the belt into the trough shape gradually. These idlers/rollers are called transition idlers/rollers (4). Similarly, transition idlers/rollers with decreasing trough angles are fixed between the last troughing idler/roller and the drive pulley to transform the belt from trough shape to flat gradually. At the feeding section, idlers/rollers with rubber rings called impact idlers/rollers (5) are installed to absorb the shock of the falling material and also to minimize the damage to the belt. An idler/roller called carrying training idler/roller (6) works to keep the belt running in the centre of the conveyor structure. These idlers/rollers are self-aligning and react to any mistracking of the belt to move into the position that will attempt to steer the belt back into the centre. The bottom strand of the belt is supported by a set of idlers/rollers called return idlers/rollers (7). Return training idler/roller (8) works similar to the carrying training idler/roller and adjust the mistracking of the return belt. The idlers/rollers are fixed to the frames, and the frames are secured to the main steel structure called stringers. Steel trusses are employed for longer-span conveyors instead of stringers. Tubular gallery structures are also used to support the belt conveyor whenever the belt conveyor is required to be enclosed. (The structure is not shown in Figure 1.1.)

Pulley is a rotating cylinder mounted on a central shaft that is used to drive, change the direction of a conveyor belt and maintain the tension in a conveyor belt. Drive pulley, also called head pulley (9), is a pulley connected to the drive mechanism of a conveyor belt. The material is discharged over the drive pulley in many conveyor systems. The pulley near the feeding end of the conveyor system is the tail pulley (10), and it turns the return run of a conveyor belt 180° back into the carrying run.

Snub pulley (11) is a small pulley often placed immediately after the drive pulley on the return side of the belt to increase the contact of the belt with the pulley, allowing the transmission of the required tension to the belt. Snub pulley is also placed before the tail pulley on the return side of the belt for improved traction. Bend pulleys (12) are the pulleys used to change the direction of the conveyor belt as in the case of the vertical take-up arrangement (16).

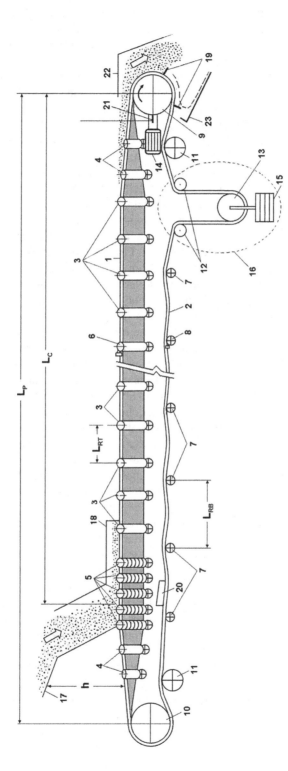

Figure 1.1 Components of a belt conveyor.

As the belt is carrying the load on it while running, the belt will stretch and slack. As a result, the belt tension will decrease. A tensioning device, called take-up pulley (13), is the pulley used to remove the slack of the belt with the help of counterweights (15) attached to it by moving while the belt is running and to maintain tension. It is installed near the drive pulley on the return side of the belt. A drive, usually an electric motor (14), is attached to the drive pulley through reduction gear and is made to rotate. As the drive pulley rotates, the belt will rotate around the drive and tail pulleys and travels linearly between the two pulleys.

The material to be conveyed is fed onto the belt by a feed chute (17) or a hopper near the tail end or at the other locations as required. The vertical or inclined plates, known as skirt plates (18), extending out from a conveyor's loading point with rubber sealing skirt material are installed closely above the belt to guide the conveyed material. This material on the carrying side is discharged in a trajectory by the head pulley.

The discharge end usually consists of discharge chute (22), belt cleaners (19), pulley cleaner (21) and a dribble chute (23). Belt cleaners remove the material which clings or adheres to the carrying or top side of the belt. Plough-type belt cleaner (20) is used to clean the bottom side of the belt on return run. Pulley cleaner removes the material adhering to the pulley. Dribble chute is an angled chute positioned under the head end of a conveyor belt to catch any material that may fall off the return side and to drop it into the discharge stream.

In addition to the above components, some other components such as skirtboard seal, wear liners, labyrinth seal, lowering chutes, trippers and ploughs are also installed in most of the conveyors depending on the requirements. Sealing system consists of elastomer seal and clamping mechanism at the edge of the skirtboard to contain dust and fines and prevent spillage. Wear liner is an abrasion-resistant plate or a layer of ceramic tiles used to line the inside of a transfer chute or skirtboard to improve material flow and prevent abrasive wear and damage to the outer shell and structure. Labyrinth seal is formed when the skirtboard and the wear liner are combined with the flexible rubber or urethane sealing strip. Lowering chutes are used at the discharge points for directing materials to storage when dusting and degradation are objectionable. Trippers and ploughs are used to discharge bulk materials from a belt conveyor at points upstream from the head pulley. Whenever the bulk material from the hopper, bin or silo is to be discharged as a feed to the belt conveyor, a piece of mechanical equipment called feeder is installed to feed the material at a regulated rate of flow.

A dust control system, called dust suppression system, using water is installed in a conveyor to reduce the escape of airborne particulates. In few conveyors, a mechanical system, called dust collection system, is installed to remove dust from the air. A weather protection system is also installed to protect the belt conveyor from the weather varying with the climate. Tramp iron detectors are used for detecting and removing the tramp iron present in the material conveyed by the belt conveyor. Continuous weighing of the bulk material conveyed by the moving conveyor belt is accomplished through the use of belt scales. Sampling devices are provided in the belt conveyor system whenever a small quantity of representative material is required for sampling from the bulk material conveyed. Safety switches are installed inevitably in all conveyors to avoid accidents.

Walkways and work platforms are also provided for ease of inspection, lubrication, clean-up and repair of the various components of the belt conveyor.

A few sensors such as belt slip sensors, belt tracking sensors and plugged chute sensors are also installed to aid in obtaining better operating conditions. Ultrasonic gauges, eddy-current-based sensors and laser-based scans are used for the measurement of conveyor cover weir. Arrangements for X-ray-based scanning and magnetic scanning of belt and belt splices are also made. Rip detectors, material spillage sensors, speed monitors and counterweight limit sensors are employed whenever necessary. All the essential components of a conveyor are sufficiently guarded for the safety of the personnel. Electrical accessories such as pre-start alarms, pull cord switches, blocked chute devices, bin level switches, misalignment switches and zero-speed switches are used in most of the belt conveyor systems to alert the personnel working in the area of belt conveyor.

1.2 Conveyor arrangements

Belt conveyors can be designed for practically any desired path of travel, limited only by the strength of the belt, angle of incline or decline, or available space. Belt conveyors may be horizontal, incline, decline, or with inclusion of concave and convex curves, or any combination of these. Figure 1.1 shows the horizontal belt. Figure 1.2 shows the typical belt conveyor travel paths.

Conveyor paths shown in Figure 1.2a and b are similar paths. However, the first conveyor is the incline or uphill conveyor where the material is transported to a higher elevation from lower elevation. The second conveyor is a decline or downhill conveyor where the material is transported to a lower elevation from higher elevation.

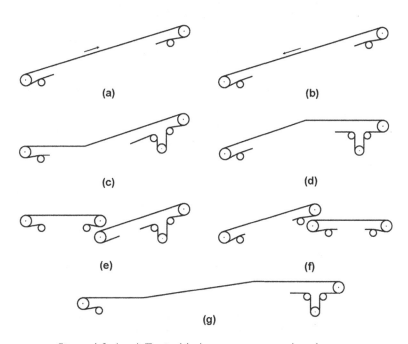

Figure 1.2 (a–g) Typical belt conveyor travel paths.

Figure 1.2c and d shows the horizontal followed by ascending path and ascending followed by horizontal path, respectively. These paths are designed when space will permit vertical curve and belt strength will permit one belt.

Figure 1.2e and f shows the horizontal followed by ascending path and ascending followed by horizontal path, respectively, with two conveyors. These paths are not preferable because of the additional wear on the belts at the feeding points, dust raised and possible plugging in the transfer chutes. Hence, paths (c) and (d) are more preferable than paths (e) and (f). Figure 1.2g illustrates a combination path that can be used on a single, long, overland conveyor. Similar paths can be used with several convex and concave curves. Overland conveyors are belt conveyors designed to carry high-tonnage loads over long distances.

Of all the main physical characteristics of the plant which will determine whether a belt conveyor can be used, the most important is the relation between the horizontal distance and the vertical distance the material is to be lifted and the flowability of the bulk material to be transported. These factors determine the angle of inclination of the conveyor. If this angle of inclination is so great that the material will roll back on the belt, specially designed belts may be necessary. Properties of bulk materials are most important to design the belt conveyor for conveying bulk materials.

Chapter 2

Properties of bulk materials

Bulk materials or **bulk solids** consist essentially of many particles or granules of different sizes (and possibly different chemical compositions and densities) randomly grouped together to form a bulk. Bulk materials are stored and handled in volumes often in unpackaged form. Examples include dry powders, granules, flakes and resins. Ores, minerals, coal, fertilizer, sulphur and salt are other common types of bulk materials.

Bulk materials are generally the most difficult state of matter to handle. Bulk materials handling is characterized by continuous-flow operations, involving materials in an aggregate form. In many cases, the bulk materials assume flow characteristics similar to those of fluids. Elements of a typical bulk materials handling system are as follows:

1. bins, silos, hoppers, bunkers for storing the material (see Section 8.1)
2. discharge devices or feeders (see Section 8.2)
3. conveyors
4. flow aid devices when needed (see Section 8.4).

The following are the most important properties of the bulk materials that affect the design of any materials handling systems.

2.1 Properties of bulk materials

2.1.1 Particle size

Various terms are in use to give a qualitative indication of the size of the particles constituting bulk materials; the word 'size' here is used loosely to mean some sort of average dimension across the particle. Table 2.1 shows the qualitative terms used to describe the size of bulk materials:

The size of the particle of standard configurations such as sphere and cube can easily be specified. For example, the size of a spherical particle is its diameter (d) and that of a cubical particle is the length of its side (l), as shown in Figure 2.1.

If the particles are of irregular shape, it is difficult to define and determine their size. Many authors have proposed several empirical definitions to particle size. Feret in 1929 defined the size of an irregular particle as the distance between two most extreme points on the surface of a particle. Martin in 1931 defined the size of an irregular

Table 2.1 Qualitative terms for various sizes

Descriptive term	Typical size range	Examples
Coarse (or broken) solid	5–100 mm	Coal, ore, aggregates, etc.
Granular solid	0.3–5 mm	Granulated sugar (0.3–0.5 mm); rice (2–3 mm)
Particulate solid:		
Coarse powder	100–300 μm	Table salt (200–300 μm)
Fine powder	10–100 μm	Icing sugar (≅45 μm)
Superfine powder	1–10 μm	Face powder
Ultrafine powder	<1 μm	Paint pigments

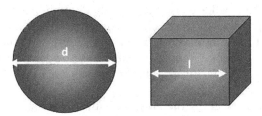

Figure 2.1 Sizes of sphere and cube.

particle as the length of the line bisecting the maximum cross-sectional area of the particle. During later years, the size of a particle is defined by comparison with a standard configuration, normally spherical particle.

In order to represent the size of an irregularly shaped particle by a single quantity, the term **equivalent size** or **equivalent diameter** is used. It is defined as the diameter of a spherical particle which behaves similar to an irregular particle under specified conditions. For instance, the diameter used could be that of a sphere which just passes through the same square sieve aperture, or which falls at the same velocity in a fluid, or which has the same projected area (microscopy).

The other two terms used to represent the size of an irregularly shaped particle are **volume diameter** and **surface diameter**.

Volume diameter, d_v, can be defined as the diameter of a sphere having the same volume as the particle.

$$d_v = \left(\frac{6V_p}{\pi} \right)^{1/3} = 1.241 \, V_p^{1/3}$$

where V_p is the volume of the particle.

Surface diameter, d_s, can be defined as the diameter of a sphere having the same surface area as the irregular particle.

$$d_s = \left(\frac{A_{sp}}{\pi} \right)^{1/2} = 0.564 \, A_{sp}^{1/2}$$

where A_{sp} is the surface area of the particle.

It is obvious that each definition has its own limitations and the usage of a particular definition of size depends on the method of measurement and the purpose for which that defined size is used. In an industrial situation, bulk materials comprise large number of particles of non-uniform sizes. In order to describe such materials completely, it is necessary to determine the particle size distribution. Particle size distribution refers to the manner in which particles are quantitatively distributed among various sizes; in other words, it is a statistical relation between quantity and size.

There are many methods for determining the particle size distribution of bulk materials. The most popular (and cheapest) method of particle size analysis, especially with relatively coarse materials, is sieving using standard test sieves. Approximate range of application of the more familiar size analysis methods and equipment is given in Table 2.2.

Particle size and particle size distribution of the material are the most important parameters that affect the bulk density and flowability of the material while handling and conveying.

2.1.2 Particle shape

Various terms, such as fibrous, angular, needle-shaped, flaky, nodular, are used to give a qualitative indication of the general shape of the particles.

In case of non-spherical particles, the term **sphericity** 'ϕ_s' is used to indicate the extent to which the particles differ from the sphere and is defined as the reciprocal of the ratio of the surface area of a particle to that of a sphere of the same volume:

$$\phi_s = \frac{\pi d_v^2}{A_{sp}} = \left(\frac{d_v}{d_s}\right)^2$$

where d_v is the volume diameter which is defined as the diameter of a sphere having the same volume as the particle and d_s is the surface diameter which is defined as the diameter of a sphere having the same surface as the particle.

The shape of the constituent particles in a material is an important characteristic as it has a significant influence on their packing and flow behaviour. Except in certain

Table 2.2 Size analysis methods

Method		Approximate range
Sieving	Dry	50 μm – 100 mm
	Wet	10 μm – 100 mm
Electrical resistance method		1 μm – 800 μm
	Coulter counter	
Laser diffraction spectrometry		1 μm – 200 μm
Sedimentation (gravity)		2 μm – 75 μm
	a. Beaker decantation	
	b. Andreasen pipette	
Elutriation		2 μm – 75 μm
Sedimentation (centrifugal)		0.05 μm – 5 μm
	Cyclosizer	
Microscope (optical)		1 μm – 150 μm
Electron microscope		0.01 μm – 1 μm

chemical processes, usually for majority of materials handling systems, the knowledge of the shape of the particles is not necessary. Most storage vessels and conveyors are designed without any consideration of the particle shape of the bulk material. However, in certain instances, the shape of the particles reveals certain behaviour. For example, a fibrous appearance could warn the tendency of the particles to lock together causing flow problems from hoppers and angular particles could be the cause of excessive wear damage to the components.

2.1.3 Surface area

The surface area is the area of the surface of the particle. The term 'specific surface' of the material is usually defined as the surface area per unit volume, but may also be defined as surface area per unit mass, depending upon the instance of using the surface area term for material.

For a single particle the volume-specific surface is thus given by

$$S_p = \frac{\pi d_s^2}{\left(\frac{\pi}{6} d_v^3\right)} \Rightarrow S_p = \frac{6}{\phi_s d_v}$$

but measurements on samples of a bulk material would, of course, yield an average value for the material. As the specific surface is inversely proportional to particle size, it is sometimes used to indicate the fineness of a powder. The most common type of instrument for measuring the surface area of powders and particulate materials is the permeameter. The surface area of certain finely divided materials such as catalysts and paint pigments is of considerable importance during the processing and further use.

2.1.4 Particle density (density of bulk solids or bulk materials)

Density of the particle is defined as the mass of the particle per unit volume. The ratio of the density of the particle to the density of water is defined as specific gravity. Bulk material is a combination of particles and space (voids). For a bulk material, the average particle density can be determined by dividing the mass of the material (solids) by the true volume occupied by the particles (not including the voids). This can be determined by using density bottle. The stepwise procedure is as follows:

1. Wash, dry and weigh the density bottle with stopper. Let this weight be w_1.
2. Thoroughly dry the ore sample.
3. Add 5–10 g of ore sample to the bottle and re-weigh. Let this weight be w_2.
4. Now fill the bottle with a liquid of known density. The liquid used should not react with the ore.
5. Insert the stopper, allow the liquid to fall out of the bottle, wipe of excess liquid, and weigh the bottle. Let this weight be w_3.
6. Remove ore and liquid from the bottle and fill the bottle with liquid alone and repeat Step 5. Let this weight be w_4.

$w_2 - w_1$ is the weight of the ore sample

$w_4 - w_1$ is the weight of the liquid occupying the whole volume of the bottle

$w_3 - w_2$ is the weight of the liquid having the volume equal to the volume of density bottle less volume of ore sample taken

$(w_4 - w_1) - (w_3 - w_2)$ is the weight of the liquid of volume equal to that of the ore sample

If ρ_1 is the density of the liquid,

$$\text{Density of the ore sample} = \frac{w_2 - w_1}{(w_4 - w_1) - (w_3 - w_2)} \times \rho_1$$

2.1.5 Bulk density

Bulk material is a combination of particles and space (voids). The percentage of the total volume not occupied by the particles is usually referred to as the 'voidage' or 'void fraction'.

$$\text{Thus, voidage, } \varepsilon = \frac{V_{\text{voids}}}{V_{\text{solids}} + V_{\text{voids}}}$$

where V_{voids} = volume of voids

V_{solids} = volume of particles.

In a bed of material having unit volume, the actual volume of solid particles, or 'fractional solids content', is $(1 - \varepsilon)$.

Sometimes the term 'porosity' is applied to bulk materials to mean the same as 'voidage'. Thus, the particle porosity can be defined as the ratio of the volume of pores within a particle to the volume of the particle (inclusive of pores).

Bulk density of the material is an apparent density and is defined as the mass of the material divided by its total volume (particles and voids).

$$\text{Thus, bulk density, } \rho_b = \frac{m_{\text{solids}} + m_{\text{voids}}}{V_{\text{solids}} + V_{\text{voids}}}$$

If ρ_p is the 'true' density of the solid particles (particle density) and ρ_f is the density of the fluid in the void spaces, then the bulk density is

$$\rho_b = (\rho_p - \rho_f)(1 - \varepsilon) + \rho_f$$

For dry bulk materials, the void spaces usually contain air and thus the density ρ_f would be negligible compared with ρ_p so that the relationship between bulk density and particle density becomes $\rho_b = \rho_p (1 - \varepsilon)$.

There are three kinds of bulk density that apply to materials handling calculations: (1) aerated density, (2) packed density and (3) dynamic or working density.

1. **Aerated density**

    When the sample of the bulk material is carefully poured into a measuring cylinder to measure its volume, then the computed density is called 'aerated', 'loose', or 'poured' bulk density (ρ_a).

2. **Packed density**

   If the sample is packed by dropping the cylinder vertically a number of times from a height of one or two centimetres onto a table, then the computed density is called 'packed' or 'tapped' bulk density (ρ_c).

3. **Dynamic** or **Working density**

   The dynamic or working density (ρ_d) is a function of aerated and packed densities. Its value is intermediate to that of aerated and packed densities. The dynamic or working density (ρ_w) is expressed as

$$\rho_d = (\rho_c - \rho_a)C + \rho_a$$

where C is the compressibility.

Bulk density of a mixture of particles of different sizes depends on the extent to which the smaller or fine particles are able to fit into the spaces among the larger ones. Hence, particle size distribution is of utmost importance in determining the bulk density. A clear knowledge of the bulk density of a material is very much essential in order to design storage vessels, conveying systems and the like.

2.1.6 Flowability of material

The flowability is determined by material characteristics such as size and shape of the fine particles and lumps, roughness or smoothness of the surface of the material particles, proportion of fines and lumps present, and moisture content of the material.

The flowability is often indicated by the compressibility of the material. Compressibility (C) is a function of packed bulk density and aerated bulk density and is expressed as

$$C(\%) = \frac{100(\rho_c - \rho_a)}{\rho_c}$$

where C = compressibility

ρ_c = packed bulk density

ρ_a = aerated bulk density.

The lower the percentage of compressibility, the material is more free-flowing. The dividing line between free-flowing (granular) and non-free-flowing (powder) is about 20%–21% compressibility. A higher percentage indicates a powder that is non-free-flowing and will be likely to bridge in a hopper. In this event, external aid of some kind is required, preferably a bin discharger. The compressibility of a material often helps to indicate uniformity in size and shape of the material, its deformability, surface area, cohesion and moisture content.

The **Carr index** is an indication of the compressibility of a powder. It is calculated by the following formula:

$$CI = \frac{V_a - V_c}{V_a} \times 100$$

where V_a is the aerated volume of a given mass of powder and V_c is the packed volume of the same mass of powder. The Carr index is frequently used in pharmaceutics as an indicator of the flowability of a powder. A Carr index greater than 25% is considered to be an indicator of poor flowability, and that below 15% indicates good flowability.

It can also be expressed as

$$CI = \left(1 - \frac{\rho_a}{\rho_c}\right) \times 100 = \left(1 - \frac{1}{H}\right) \times 100$$

where ρ_a is the aerated density, ρ_c is the packed density of the powder, and H is the **Hausner ratio** or **flow ratio** defined as the ratio of packed density to the aerated density.

2.1.7 Hardness

Hardness is the characteristic of a solid material expressing its resistance to permanent deformation. Hardness can be measured on the Mohs scale or various other scales. Mohs based the scale on ten minerals that are all readily available. The Mohs scale of hardness is shown in Table 2.3.

The hardness of a material is measured against the scale by finding the hardest material that the given material can scratch, and/or the softest material that can scratch the given material. For example, if some material is scratched by apatite but not by fluorite, its hardness on the Mohs scale would fall between 4 and 5. Hardness provides an indication of the abrasive wear rate encountered in bins, hoppers, chutes, gates and other pieces of equipment.

2.1.8 Cohesion and adhesion

Cohesion is defined as the molecular attraction by which particles of a body or material are united or held together. When the forces of attraction (cohesion) are low, the bulk material can be made to flow easily under the influence of gravity with the particles moving as individuals relative to one another. Dry sand and granulated sugar are the familiar examples of free-flowing bulk solids. However, high inter-particle cohesive forces, which may be caused by moisture or electrostatic charging, result in the formation of agglomerates so that the material flows in an erratic manner as lumps. Wheat flour is an example for cohesive bulk solids which usually exhibit this sort of behaviour.

Table 2.3 Mohs scale of hardness

Hardness	Mineral	Chemical formula	Explanation
1	Talc	$Mg_3 Si_4 O_{10} (OH)_2$	Very soft, can be powdered with the finger
2	Gypsum	$CaSO_4 \cdot 2H_2O$	Moderately soft, can scratch lead
3	Calcite	$CaCO_3$	Can scratch fingernail
4	Fluorite	CaF_2	Can scratch a copper coin
5	Apatite	$Ca_5 (PO_4)_3 (Cl, F)$	Can scratch a knife blade with difficulty
6	Feldspar	$KAlSi_3O_8$	Can scratch a knife blade
7	Quartz	SiO_2	All products harder than 6 will scratch window glass
8	Topaz	$Al_2 F_2 SiO_4$	
9	Corundum	Al_2O_3	
10	Diamond	C	

A direct measurement of cohesion is difficult and requires sophisticated equipment such as a 'cohetester'. An alternative measurement can be made of the material's **uniformity coefficient**. This is a numerical value obtained by dividing two factors: the size of a sieve opening that will pass 60% of the sample divided by the size of the sieve opening that will pass 10% of the sample. The flowability of a material is directly affected by its uniformity in both size and shape. Obviously, the greater the diversity of the material in size and shape, the less readily it will flow.

In addition to indicating particle size and shape, the uniformity coefficient of a material also provides an indication of its compressibility. The more compressible the material is, the more the trouble it will be to move it through the various stages of a process. Generally, if a material is not compressible, it will flow readily.

Adhesion is the sticking together or adhering of substances in contact with each other. The distinction between adhesion and cohesion is sometimes blurred; cohesion is internal, whereas adhesion is external. When designing systems involving the flow of bulk solids from hoppers or in chutes, or in any situation where a bulk solid slides in contact with a fixed boundary surface, the property of adhesion is important. Adhesion describes the tendency of solid particles to 'stick' to a containing surface, such as a wall of a hopper, the side and bottom surfaces of a chute or the surface of a conveyor belt. Kaolin clay is an extreme example for this type of material, which is so tacky that it will stick to a wall when thrown against it. This can create unusual problems in moving this material from storage. Adhesive materials tend to bridge in storage and thus require external assistance.

2.1.9 Angle of repose

It is defined as the included angle formed between the edge of a cone-shaped pile formed by dropping the material from a given elevation (6.8 cm – according to the standard) and the horizontal. This angle depends principally upon the nature of the bulk solids.

Actually, there are two separate types of angle of repose: filling (poured angle of repose) which occurs when a material is poured into bins, trucks, railcars or other storage structures such as stock piles, and emptying (drained angle of repose) when it is removed, or drained, from them (Figure 2.2). Filling and emptying angles of repose are not always equal.

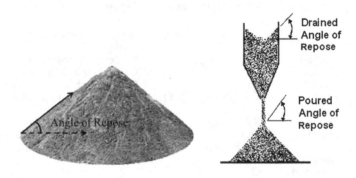

Figure 2.2 Poured and drained angles of repose.

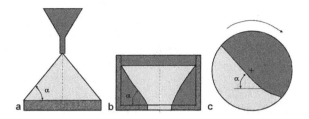

Figure 2.3 Methods to measure the angle of repose.

In general, it is related to many of the flow properties of a material and is thus an indirect indication of the flowability of the material. Many methods have been devised for measuring the angle of repose of bulk solids. Figure 2.3 illustrates three such methods.

The most commonly used method yields a value of poured angle of repose, which is the angle between the horizontal and the sloping side of a heap of the material poured gently from a funnel onto a flat surface (a). Another method involves the determination of drained angle of repose with the bulk solid flowing out of a container with a central outlet (b). In a rotating drum, another (smaller) angle of repose is obtained (c).

The measurement of angle of repose gives a direct indication of how free-flowing a material will be. A low angle of repose indicates that the material flows readily, and a high angle of repose indicates that it does not flow. Generally, it is safe to treat the angle of repose as an only indicator of the contours of heaps of the material. Table 2.4 shows the flowability character based on the angle of repose.

2.1.10 Angle of fall

When a material lies in a pile at rest, it has a specific angle of repose. If the supporting surface experiences vibrations, impacts or other movements, the material on the sloped sides of the pile will dislodge and flow down the slope. The new angle of repose that forms is called the angle of fall. After measuring the angle of repose, the cone-shaped pile of material is jarred by dropping a weight near it. The pile will fall resulting in a new, shallower angle with the horizontal. This new angle of repose is measured as angle of fall. The way the pile falls is of special interest. If particles fall and spread out along the slope of the pile, only the degree of flowability is indicated. If the entire pile collapses, it indicates that the material contained entrained air and is prone to flushing. This parameter provides an indication of particle size, shape, uniformity, cohesion and thus the flowability of the material.

Table 2.4 Flowability character based on angle of repose

Angle of repose	Flow character
25°–30°	Very free-flowing
30°–38°	Free-flowing
38°–45°	Fair flowing
45°–55°	Cohesive
>55°	Very cohesive

2.1.11 Angle of difference

Angle of difference is the difference between the angle of repose and the angle of fall. The greater the angle of difference (between angles of repose and fall), the more free-flowing is the material. It is an indirect measure of fluidity, surface area and cohesion.

2.1.12 Angle of spatula

When a flat blade is inserted into a pile of granular material and is lifted vertically, a new angle will form with the material relative to the blade surface. This angle is known as the angle of spatula. Generally, bulk solids with an angle of spatula less than approximately 40° are considered free-flowing. A highly flowable material will have an obtuse angle of spatula. The angle of spatula provides an indication of the internal friction between particles. It is an indirect measurement of cohesion, surface area, size, shape, uniformity, fluidity, deformability and porosity of the material.

2.1.13 Angle of slide

This is the angle of a flat surface with the horizontal at which a material will slide down a flat surface due to its own weight. Angle of slide provides an indication of flowability of the material and is particularly useful in hopper and chute design. In case of incline or decline conveyors, the inclination of the belt conveyor should always be less than the angle of slide.

2.1.14 Angle of surcharge

It is defined as the included angle formed between the edge of a cone-shaped material and the horizontal when the material is at rest on a moving surface such as conveyor belt. This angle is usually 5° to 15° less than the angle of repose though in some materials it may be as much as 20° less. The angle of surcharge is often called dynamic angle of repose (Figure 2.4).

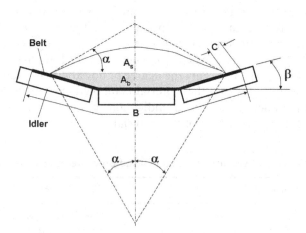

Figure 2.4 Cross-sectional area of the material on a conveyor belt. B = Belt width; A_s = area of surcharge; A_b = area of base trapezoid; α = angle of surcharge; β = troughing angle; C = free edge distance.

The flowability of the material, as measured by its angle of repose and angle of surcharge, determines the cross section of the material load which can be safely carried on a belt. It is also an index of the safe angle of incline of the belt conveyor. Table 2.5 illustrates and defines the normal relationship and general characteristics of materials.

2.1.15 Abrasion

Abrasion means scraping off or wearing away. Knowing a material's abrasiveness is important to properly design the equipment to protect against wear. Abrasiveness is a composite of several different material characteristics, including strength, hardness and particle shape. Abrasive wear rate varies directly with the pressure and velocity of the particle movement across a surface. Chute and hopper slopes should be designed as close as possible to the angle of slide of the material. Steeper slopes move material, but result in greater wear. Hardened steels, wear-resistant liners and high-density plastics must be considered for contact materials. Material dead beds at impact points reduce wear, but may induce another problem of tendency to build up into an immobile mass.

2.1.16 Corrosion

Corrosion means the deterioration in the material and its properties due to chemical or electrochemical reaction between a material, usually a metal, and its environment. This means a loss of electrons of metals reacting with water and oxygen. Weakening of iron due to oxidation of the iron atoms is a well-known example of electrochemical corrosion. This is commonly known as rust. Most structural alloys corrode merely from exposure to moisture in the air, but the process can be strongly affected by exposure to certain substances. Corrosion can be concentrated locally to form a pit or crack, or it can extend across a wide area to produce general deterioration.

Corrosiveness usually is defined by the pH of the material. A material with a pH of 1–5 is very corrosive, a material with a pH of 5–7 is mildly corrosive, and a material with pH greater than 7 is generally non-corrosive. When corrosive materials are processed, they must be handled in equipment with contact surfaces of alloy steel, special plastics or coatings with corrosion-resistant paint. When abrasion is also present, the effectiveness of coatings is reduced.

2.1.17 Friability (degradation)

This quality can usually be determined by a simple squeeze test, similar to that used in checking compressibility. If it is undesirable to have any breakdown of the product in the process, as in food products such as cereals or vegetables or as in coal, it is mandatory to use equipment whose design or performance will prevent such breakage.

2.1.18 Dispersibility or dustiness

Dispersibility of a powder is its propensity to emit dust when undergoing stress in a process or handling operation. Dispersibility is the basic property that causes a material to flood or to produce dustiness in the surroundings. Dispersibility indicates the dusting and flushing characteristics of a material. It is a measure of the propensity

Table 2.5 Relationship between flowability, angle of repose and surcharge

Very free-flowing	Free-flowing	Average flowing		Sluggish
5° Angle of surcharge	10° Angle of surcharge	20° Angle of surcharge	25° Angle of surcharge	30° Angle of surcharge
0° – 19° Angle of repose	20° – 29° Angle of repose	30° – 34° Angle of repose	35° – 39° Angle of repose	40° and up Angle of repose
Uniform size, very small rounded particle, either very wet or very dry, such as dry silica sand, cement, wet concrete, etc.	Rounded, dry polished particles, or medium weight, such as whole grain and beans, etc.	Irregular, granular, or lumpy materials of medium weight, such as anthracite coal, cottonseed meals, clay, etc.	Typical common materials such as bituminous coal, stone, most ores, etc.	Irregular, stringy, fibrous, interlocking materials such as wood chips, bagasse, tempered foundry sand, etc.

Courtesy: CEMA.

for a granular material to form dust and thus lose mass to the surrounding air. It is especially important during vertical flow (e.g. pile formation after spout or conveyor discharge). The more dispersible the material is, the higher the potential for mass loss due to dust generation. Dispersible materials are generally of low bulk density and fine particle size, which causes them to behave more like a gas or a liquid than a solid. Dispersibility and floodability are interrelated. Materials with a dispersibility rating of more than 50% are very floodable and are likely to flush from a storage bin unless measures are taken to prevent this occurrence.

2.1.19 Flowability index

A composite evaluation of the flow characteristics of a granular material involves the use of angle of repose, compressibility, angle of spatula and uniformity coefficient. After measuring the individual values of these four properties, each is assigned an index value according to Carr's procedure. The sum of the individual indices constitutes the 'flowability index' of the granular material. A maximum flowability index value of 100 indicates that material flowability is very good, while a low index indicates the material flowability is very bad. The flowability index also partly indicates the potential floodability of a bulk material. The higher the flowability index, the higher the risk that the material may flood.

2.1.20 Floodability index

Floodability of a material is its tendency to have a liquid-like flow characteristic, which is caused by aeration within the powder. Powders can, when discharged from a hopper, behave in an unstable, liquid-like manner, which may be discontinuous, gushing and uncontrollable. The reasons for flooding are generally due to excessive aeration within the powder bed, which leads to easy powder fluidization and ultimately to flooding.

The evaluation of the potential floodability of a material involves the use of the flowability index, as well as the measured values of angle of fall, angle of difference, and dispersibility. The sum of these four parameters provides the floodability index of bulk material. A maximum index value of 100 indicates that the chance of floodability is very high, whereas a low value indicates that the material should not flush.

2.1.21 Permeability

A powder is a biphasic system. The ability of a gas to pass through the particles of powder is called permeability. The magnitude of permeability as a function of bulk density may be used in the determination of the powder flow rate and discharge of powders from hoppers.

2.1.22 Moisture content and hygroscopicity

Moisture in the material causes effects such as chemical change, deterioration in quality. Moisture also has a dramatic influence on the flow behaviour of bulk solids. Therefore, moisture analysis is one of the most frequently performed tasks in their characterization. The moisture content is usually expressed in terms of the percentage moisture in the material. Thus,

$$\text{Percent moisture} = \frac{\text{Weight of water}}{\text{Weight of solids and water}} \times 100$$

A simple method to determine the moisture content is to weigh a sample of the material and then place the sample in an oven for an appropriate length of time to dry it thoroughly. Then the percent moisture is calculated from

$$\text{Percent moisture} = \frac{\text{Initial weight of sample} - \text{final weight of sample}}{\text{Initial weight of sample}} \times 100$$

Free moisture, surface moisture or combined moisture as in sludges or wet ores causes the classic problem of sticking and poor flow. Generally, free moisture over 5%–10% is considered risky. Likewise, those materials that are hygroscopic or absorb moisture are almost certainly a problem. When storage bins are not covered and are, therefore, exposed to weather, materials such as salt will cake and refuse to flow. Others such as sugar and lime will cake and flow erratically or will not flow at all in conditions of high humidity. If it is the only factor, this condition is amenable to several remedies like keeping the storage bin covered, keeping the system enclosed, air conditioning or moving the plant to a dry climate.

2.1.23 Static charges

Some materials are subject to a build-up of static electricity which can cause sticking and clogging in storage and processing equipment. High-volume production, requiring high feed rates, can generate a static build-up of sufficient magnitude which requires some means of circumventing the problem. The usual remedy is a static inhibitor coating or grounding strap.

2.1.24 Temperature limits

In the plastic industry, some materials are stored when they are hot and then allowed to cool. As the hot material cools, the particles may adhere and form agglomerates, with the result that flow characteristics are adversely affected. Materials with low melting points are most affected. Other materials such as dry cement and gypsum become extremely fluidizable when hot, radically altering their low-temperature behaviour.

2.1.25 Explosiveness

Many bulk materials when dispersed in air to form a dust cloud or suspension constitute a potentially explosive mixture which may be ignited by a naked flame, a hot surface or an electrical discharge and rapidly propagate a flame through the suspension, with a subsequent sudden increase in pressure as a result of the release of heat and gaseous products from the burning dust. This is commonly called a **dust explosion**. The range of products that are hazardous in this respect is quite wide and includes common foodstuffs such as sugar, flour and cocoa; synthetic materials such as plastics, chemicals and pharmaceuticals; metals such as aluminium and magnesium; and traditional fuels such as coal and wood. Other products such as sand, alumina and certain paint pigments are non-combustible and therefore present no danger.

Explosion hazard is caused by only the fraction of the bulk material having a particle size less than about 200 μm. Oxidation of such fine particles occurs rapidly, in association with a rapid rise in temperature since the surface area of the particles in contact with the air is large and their volume is relatively small. The risk of an explosion to occur depends upon the presence of the following three conditions:

1. a suspension of combustible dust of explosive concentration
2. an ignition source
3. oxygen in sufficient quantity to support combustion.

If any one of these conditions does not exist, it will be impossible for an explosion to occur. If dust clouds and sources of ignition are eliminated, explosion hazard will not occur. If the risk is still considered to exist, the use of inert gas avoids the hazard.

It is essential to collect a representative sample from the bulk material for testing to measure the required properties before the material is handled, conveyed, transported or stacked. Description of different sampling methods and their applicability is out of the scope of this book. Hence, standard books on sampling methods are to be referred for standard procedures to collect representative samples.

2.2 Behaviour of materials on a moving belt

Characteristics of the materials are influenced considerably by the movement, slope and speed of the conveyor belt that carries the material. When the material is carried by the conveyor belt, the material is agitated while the conveyor belt passes successively over each carrying idler. This agitation tends to make the larger pieces at the top of the material with the smaller particles or fines at the bottom. Agitation also tends to flatten the surface slope of the material, and thus the angle of surcharge is less than the angle of repose. When the material is fed onto a conveyor belt at a velocity not equal to the velocity of the conveyor belt, there would be turbulence in the material. Further turbulence is produced if the vertical velocity of the material when loading the belt is not absorbed by conveyor belt and impact idlers. These influences are more pronounced in incline or decline conveyors, when conveyor belt is operated at high speeds, and the material handled is loose and contains large rounded lumps.

Chapter 3

Belt tension

In any belt drive, whether it is for a transmission or a conveyor, there exists a difference in tension in the belt on the two sides of the drive pulley. The larger tension, which is at the carrying run as the discharge pulley is the drive pulley, is called the **tight side tension** (T_1) and the smaller, which is at the return run, is called the **slack side tension** (T_2). Without slack side tension the belt cannot be driven. The difference between the tight side and slack side tensions is known as the **effective tension** (T_E). This is the tension which actually does the work.

The horsepower (HP) required at the drive of a belt conveyor is derived from the pounds of effective tension required at the drive pulley to propel or restrain the loaded conveyor at the design velocity of the belt V, in ft/min:

$$HP = \frac{T_E V}{33,000}$$

The power required to drive the belt is a function of effective tension produced by the resistance forces. The overall resistance is made up of the following resistances:

Main resistances

- friction in bearings and seals in idlers
- due to the movement of the belt from indentation of the belt by idlers, flexing of the belt and movement of the material

Secondary resistances

- inertial and frictional resistances due to the acceleration of the load in the loading area
- friction on the side walls of chutes in the loading area
- friction of pulley bearings (not driving pulleys)
- resistance due to wrapping of the belt round pulleys

Special main resistances

- drag due to forward idler tilt
- friction from chute flaps or skirts over the full belt length

Special secondary resistances

- friction from belt and pulley cleaners
- friction from chute flaps or skirts over the part belt length
- resistance due to inverting the return strand of the belt
- resistance due to discharge ploughs
- resistance due to trippers

Slope resistance

- due to the lifting or lowering of the load in incline or decline conveyors

As the determination of these resistances is complex, the effective tension can be considered as the summation of the belt tensions produced by forces such as

1. the gravitational load to lift or lower the material being transported
2. the frictional resistance of the conveyor components, drive, and all accessories while operating at design capacity
3. the frictional resistance of the material as it is being conveyed
4. the force required to accelerate the material continuously as it is fed onto the conveyor by a chute or a feeder.

The basic formula for calculating the effective belt tension at drive, T_E, is

$$T_E = LK_t \left(K_x + K_y W_b + 0.015 W_b \right) + W_m \left(LK_y \pm H \right) + T_p + T_{am} + T_{ac}$$

where

L = the length of the conveyor, ft
H = the vertical distance that the material is lifted or lowered, ft
K_t = the ambient temperature correction factor (dimensionless)
K_x = the factor used to calculate the frictional resistance of the idlers and the sliding resistance between the belt and idler rolls per unit length, lbs/ft
K_y = the factor used to calculate the resistance of the belt and the load to flexure as the belt and load move over the idlers (dimensionless)
W_b = the weight of the belt, pounds per foot of belt length
W_m = the weight of the material, pounds per foot of belt length
T_p = the tension resulting from the resistance of the belt to flexure around pulleys and bearings, total for all pulleys, lbs
T_{am} = the tension resulting from the force to accelerate the material continuously as it is fed onto the belt, lbs
T_{ac} = the total of the tensions from conveyor accessories, lbs

$$T_{ac} = T_{sb} + T_{pl} + T_{tr} + T_{bc}$$

T_{sb} = the tension resulting from the force to overcome skirtboard friction, lbs
T_{pl} = the tension resulting from the frictional resistance of ploughs, lbs
T_{tr} = the tension resulting from the additional frictional resistance of the pulleys and the flexure of the belt over units such as trippers, lbs
T_{bc} = the tension resulting from the belt pull required for belt cleaning devices such as belt scrapers, lbs

The effective tension may be calculated from wattmeter power readings of actual installations. In this case, motor and drive losses must be deducted from the electrical input to the motor, leaving the power absorbed by the belt. Substituting this latter value in the foregoing equation, with the proper value of belt speed, also provides the effective tension. It should be noted that drive and motor losses must be added to the electrical output of the motor (generator) to obtain the belt horsepower in the case of decline belts requiring restraint.

Usually, the slack tension is obtained by a counterweight or by a screw-type take-up. The former is preferable, because it maintains a constant tension automatically and may be set at the lowest tension at which the conveyor can be driven. This type maintains a constant tension under all conditions of load, starting, stretch, and so on.

The force required to drive a belt conveyor must be transmitted from the drive pulley to the belt by means of friction between their two surfaces. The force required to restrain a downhill regenerative conveyor is transmitted in exactly the same manner. In order to transmit power, there must be a difference in the tensions in the belt as it approaches and leaves the drive pulley. This difference in tensions is supplied by the driving power source.

If power is transmitted from the pulley to the belt, the approaching portion of the belt will have the larger tension (T_1) and the departing portion will have the smaller tension (T_2). If power is transmitted from the belt to the pulley, as with a regenerative decline conveyor, the reverse is true. Regenerative conveyor is a conveyor that conveys material downhill and discharges at a substantially lower altitude than the tail generating power rather than demanding power for its operation. This means it will tend to run on its own when adequately loaded. Figure 3.1 illustrates the typical arrangements of single-pulley drives in incline and decline conveyors.

3.1 Wrap factor, C_W

Wrap is used to refer to the angle or arc of contact the belt makes with the pulley's circumference. The wrap factor, C_W, is a mathematical value used in the determination of the effective belt tension, T_E, that can be dependently developed by the drive pulley. The effective belt tension is governed by the coefficient of friction existing between the pulley and the belt, wrap and the values of T_1 and T_2.

The following are the formulae used to evaluate the drive–pulley relationships:

$$\frac{T_1}{T_2} = e^{f\theta} = \frac{1+C_W}{C_W}; \quad C_W = \frac{T_2}{T_E} = \frac{1}{e^{f\theta}-1}$$

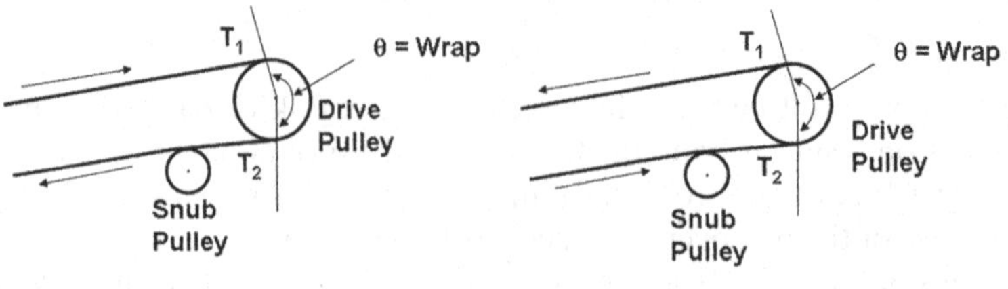

Figure 3.1 Single-pulley drives in incline and decline conveyors.

where

C_W = the wrap factor
T_1 = the tight side tension at pulley, lbs
T_2 = the slack side tension at pulley, lbs
e = the base of Napierian logarithms = 2.718
f = the coefficient of friction between pulley surface and belt surface (0.25 for bare pulley and 0.35 for lagged pulley)
θ = the wrap of belt around the pulley (or) wrap angle in radians (one degree = 0.0174 radians)

$$T_E = T_1 - T_2 = \text{effective belt tension, lbs}$$

Wrap angle (arc of contact), θ, is defined as the angle subtended by the belt on pulley. The angle of wrap of the belt around the drive pulley can be varied by the use of a snub pulley as shown in Figure 3.2.

The wrap factor does not determine T_2, but establishes its safe minimum value for a dry belt. A wet belt and pulley will substantially reduce the power which can be transmitted from one to the other because of the lower coefficient of friction of the wet surfaces. The following are some of the remedies to increase the transmission of power:

1. groove the lagging on the pulley
2. keep the driving side of the belt dry
3. increase the wrap angle
4. increase the slack side tension by increasing counterweight in a gravity take-up.

The coefficient of friction may vary when driving a rubber-surfaced belt by a bare steel or cast iron pulley, or by a rubber-lagged pulley surface. For most cases, the belt will have an angle of wrap around the drive pulley of about 180° to 240°.

3.2 Drive arrangements

There are many drive arrangements that can be constructed depending on the general conveyor profile. In a single-pulley drive arrangement, the drive for a belt conveyor may be at either end of the conveyor, although it is generally better to drive the head pulley as shown in Figure 3.3a and b as this will involve the smallest amount of belt being subjected to maximum tension. In a single-pulley drive without snub pulley, the

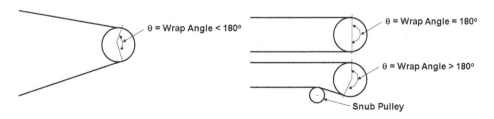

Figure 3.2 Increase in wrap angle using snub pulley.

angle of wrap is limited to 180°. The snubbed drive is more popular since in addition to the larger angle of wrap it has the advantage that it brings the carrying and return strands of the belt closer together and thus reduces the depth of the supporting struc- ture required. An alternative arrangement is to have the drive at an intermediate point on the return strand of the belt close to the head pulley as shown in Figure 3.3c. Where the conveyor is operating downhill and involves regenerative effects, the tail pulley should be driven as in Figure 3.3d or a separately driven pulley on the return strand is fitted as close as possible to the tail end as in Figure 3.3e. With a tandem drive, two pulleys are geared together and driven by a single motor as in Figure 3.3f, and this arrangement gives an angle of wrap almost double that of a single pulley.

The effectiveness of the conveyor drive is dependent upon a number of factors, prin- cipally the effective tension, the friction between the belt and the drive pulley and the angle of wrap, or arc of contact, of the belt to the pulley. The power that can be

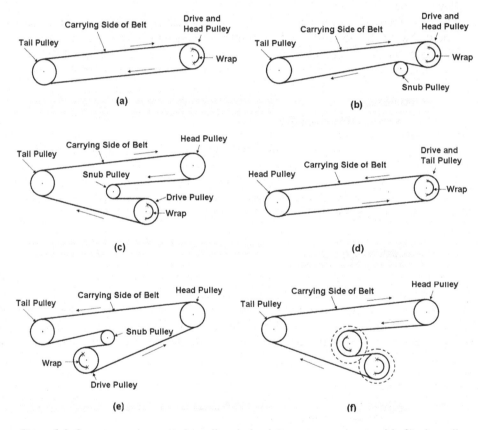

Figure 3.3 Some common single-pulley belt drive arrangements. (a) Single-pulley drive at head end of conveyor without snub pulley, (b) single-pulley drive at head end of conveyor with snub pulley, (c) single-pulley drive on return run, (d) single-pulley drive at tail end without snub pulley, regenerative, (e) single-pulley drive on return run, regenerative, (f) tandem drive on return strand.

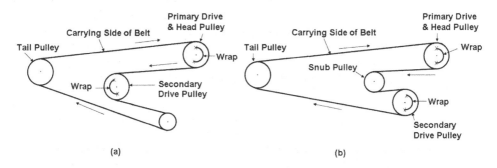

Figure 3.4 Dual-pulley drive arrangements. (a) Dual-pulley drive, (b) dual-pulley drive having both pulleys in contact with the clean side of the belt.

transmitted from the driving pulley to the belt is limited by the point at which the belt begins to slip. In order to increase the power, it is necessary either to increase the coefficient of friction, for example by applying a rubber lagging to the surface of the pulley, or to increase the angle of wrap by snubbing the pulley or providing a dual drive.

Dual-pulley drive arrangement uses two or more separate motors, one or more driving the primary drive pulley and one or more driving the secondary drive pulley. When a dual-pulley drive as in Figure 3.4a is used, there is a major increase in wrap as in the case of tandem drive. This increase in available wrap can often provide for lower maximum belt tension and a more efficient and lower-cost conveyor design. A drawback in tandem as well as dual-pulley drive arrangements is that one of the driving pulleys will be in contact with the carrying (i.e. dirty) side of the belt. Another disadvantage of the geared tandem drive is that, because of slight differences in the tension of the belt as it passes over the pulleys, there will inevitably be some slip between the belt and the second pulley. Dual-pulley arrangement as in Figure 3.4b is employed to make only the clean side of the belt be in contact with the two pulleys. The design of a conveyor drive arrangement is influenced by many factors, including the performance requirements, the preferred physical location and relative costs of components and installation.

3.3 Maximum and minimum belt tensions

Maximum belt tension occurs when the belt is conveying the design load from the feeding point continuously to the point of discharge. It usually occurs at the discharge point on horizontal or incline conveyors and at the feeding point on regenerative decline conveyors. In compound conveyors with a profile containing an incline, a decline, and then another incline, higher tension may be generated temporarily when only the inclines are loaded and the decline is empty. The location and magnitude of the high tension must be determined for selecting the belt.

Maximum belt tension also occurs during starting and stopping of the belt conveyor. As the starting torque of an electric motor is high, starting belt tension will be high as the motor has to transmit high torque to a conveyor belt. During stopping, when the belt is brought to rest very rapidly, especially on decline conveyors, the inertia of the loaded belt may produce high tensions. These temporary starting and stopping tensions must be avoided to prevent weakening and/or failure of belt splices.

For conveyors that do not overhaul the drive, the minimum belt tension on the carrying run will usually occur at the tail (feed) end. For conveyors that do overhaul their drive, the minimum belt tension will usually occur at the head (discharge) end.

Chapter 4

Conveyor belt

Conveyor belt is the most important and costliest part of the belt conveyor. Since the conveyor belt is the most vulnerable part of the conveyor, the selection of the conveyor belt must be made with meticulous care.

4.1 Parts of conveyor belt

The conveyor belt works as a tractive component as well as load-carrying component in belt conveyor. It may be used for transportation of different kinds of material at a higher speed (6–8 m/sec). For this purpose, the conveyor belt needs to have the following essential properties:

1. High strength
2. Low mass per unit length
3. Low stretch or elongation under normal working stresses
4. Transverse rigidity
5. Flexibility
6. Simplicity and inexpensive
7. Longer life
8. Wear resistance
9. Fire resistance

A conveyor belt (Figure 4.1) is generally composed of two main integral parts:

1. **Carcass** – It is the reinforcement inside of a conveyor belt. It may be ply type as in fabric belts or steel cord (or cable) type, which provides sufficient strength to handle the operating tensions and to support the load.

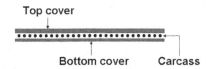

Figure 4.1 Cross section of the conveyor belt.

2. **Top and bottom covers** – They are also called conveying and pulley covers constructed of different rubber compounds with varying physical properties and chemical resistance to protect the carcass and give the conveyor belt an economical life span.

Cotton fibre reinforcement for carcasses and natural rubber for covers were used in the past for conveyor belt. A broad range of cotton and synthetic fibres for carcass, synthetic rubbers, polymers and elastomers for top and bottom covers have been developed to increase the service life and to extend the operational range of the conveyor belt. Hence, a wide range of types and constructions of conveyor belt are available today.

4.2 Belt carcass

The belt carcass is the primary reinforcement and most important structural unit of the belt. It is the heart of the conveyor belt. It must be capable of withstanding all the stresses developed in the belt as it receives and carries the load. The carcass must have the following:

- Adequate tensile strength to withstand the maximum belt operating tension and to move the loaded belt
- Sufficient longitudinal flexibility for good operation over pulleys
- Sufficient transverse flexibility for proper troughing when running empty
- Sufficient longitudinal and transverse stiffness required to the belt for the load support as the belt is moving on spaced idlers
- Ability to absorb impact of the material as it is loaded on the conveyor belt
- Ability to transport the load smoothly and without spillage
- Ability to operate at temperatures in excess of 90°C
- Low self-weight and small specific elongation at working tension
- Good tear and rip resistance to withstand damage
- Good mechanical fastener holding ability
- Good adhesion between plies
- High resistivity to ply separation
- Low hygroscopicity and long service life

The carcass is designated by the working tension (newtons per millimetre of width or kilonewtons per metre of width or pounds per inch of width) or the strength of the carcass. There are two separate strength measurements for the carcass or the belt. The first one is the **maximum working tension** or strength of the belt. This is the highest tension occurring in any portion of the belt on the conveyor system under normal operating conditions. This is the strength measurement used to determine the proper belt for the system. The second measurement is the **ultimate tensile strength** of the belt. It is the point at which the belt will rupture and fail due to excessive tension. The difference between the maximum working tension and the ultimate tensile strength of the belt is often referred to as the **service factor**. For top-quality belt, this service factor is 10 to 1. This means that if the maximum working tension is 200 PIW (pounds per inch of width), the ultimate tensile strength would be 2,000 PIW. The carcass is normally rated by the manufacturer in terms of "maximum recommended operating tension" permissible.

A fibre is a unit of matter having a length at least 100 times its diameter and which can be spun into a yarn. **Filament fibres** refer to fibres of long continuous lengths, while **staple fibres** refer to those of shorter lengths, which are about a few inches long. Most natural fibres, such as cotton and wool, are staple fibres. Synthetic fibres, such as nylon and polyester, are considered filament fibres. The natural fibre silk is also a filament fibre, but when filament fibres are cut short, they are considered staple fibres.

Yarn is a generic term for a continuous strand of textile fibres, filaments or material in a form suitable for knitting, weaving or otherwise intertwining to form a textile fabric. Yarns made from short, staple fibres are called **spun yarns**. Yarns made from continuous filament fibres are called **filament yarns**. Filament yarns are stronger than the same-size spun yarns of the same synthetic material. Textile fabric belts are the most commonly used belts for conveyors.

Most conveyor belt carcasses are made of one or more plies of woven fabric. A **ply** is a reinforcement of fibre weaved in a definite pattern which latter impregnated with rubber and protected by rubber cover to make the complete belt. The simplest type of fibre or yarn is called **singles** yarn. Singles yarn consists of a number of very small fibres bound together and usually twisted somewhat to make the yarn behave as a unit. When two or more of these singles yarns are twisted together, the result is **plied** yarn. If two or more plied yarns are twisted together, the result is **cable** or **cord**. The yarns of required size and strength are used to make a ply, which appears similar to cloth/mat, but is very strong, tough and flexible to suit tension rating of carcass/belt.

Conveyor belt fabric is made of two yarns weaved usually at right angles to each other. The yarns that run lengthwise (parallel to the conveyor) are referred to as **warp yarns** and are the tension-bearing members. The transverse or cross yarns are called **weft yarns** or **fillings** and are primarily designed for impact resistance and general fabric stability.

Three main types of commonly used weave patterns are **plain weave**, **straight warp weave** and **solid-woven weave** and are shown in Figure 4.2.

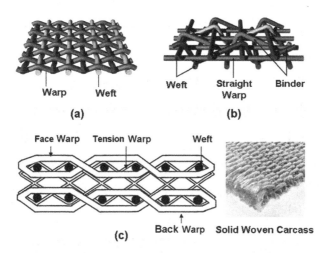

Figure 4.2 Three main types of belt weaves. (a) Plain weave, (b) straight warp weave and (c) solid-woven weave. (Courtesy – Fenner Dunlop Engineered Conveyor Solutions.)

Plain weave pattern is the oldest and most common type of belt weave, in which the warp and weft yarns cross each other alternately. In some cases, warp or weft yarns appear in pairs rather than as singles yarns. Belts utilizing this weave are popular because of the simple fabric construction leading to a relatively inexpensive belt, which can be easily spliced.

Straight warp weave has straight-laid, non-interwoven warp yarns, and straight-laid, non-interwoven weft yarns, plus binder warp yarns which are interwoven with the weft yarns to hold the structure together. The straight-laid warp yarns are the primary tension element in this type of weave. Straight warp fabric provides a belt that delivers excellent load support – up to three times greater impact resistance than traditional plied belts and five times the longitudinal rip resistance of the equivalent rated multi-ply construction. This type of weave fabric is stronger than plain weave fabric. Straight warp fabric belts are used in higher tension applications. They are more expensive and more difficult to splice.

Solid-woven weave is a multiple weave, with at least two warp systems and two or more planes of weft yarns. The multiple layers of warp and weft are tied together with a binder yarn. This weave can be considered as an extension of the straight warp concept. As in straight warp weave, the yarns employed are much thicker in this weave. Solid-woven weave gives the belt exceptional impact and tear resistance and is ideal for accepting mechanical splices. This type of weave is primarily used in single-ply high-tension belts.

When woven textile fabric is used for reinforcement, warp yarns are required to have a high tensile strength and low elasticity (elongation), whereas weft yarns must have good elasticity and energy absorption capacity. Some types of weave patterns are plain, rib, basket, twill, crowfoot, satin, crowfoot satin and leno. These weave patterns are shown in Figure 4.3.

The plain weave is the basic woven fabric structure and is very widely used. It forms the basis for the development of the more complicated interlacing patterns

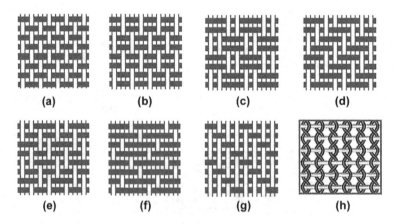

Figure 4.3 Types of belt weave patterns. (a) Plain, (b) rib, (c) basket, (d) 2/2 twill, (e) crowfoot, (f) satin, (g) crowfoot satin and (h) leno.

used in weaving. Rib weave is a variation of a plain weave. Either warp or weft yarn is thicker than the other. Usually, warp yarn is thicker, in which case there is an appearance of prominent ribs in the horizontal direction.

Basket weave is highly similar to plain weave with the same weaving but with two or more yarns combined and woven as one. The advantage is that by running more than one thread together, the tear strength of the fabric is improved. This simple modification to the basic weave gives a most significant improvement to the tear resistance of the final fabric compared with the basic plain weave.

In twill weave, warp and weft threads still pass alternately over two and under two, but instead of each pair of threads running together, each is displaced one thread from the adjacent thread, thus giving a pronounced diagonal appearance to the fabric. In this, weave pattern repeats every four threads in both warp and weft. Hence, twill weave has a characteristic diagonal rib formed by the interlace points in the fabric.

With both the basket and twill weaves, the longer 'float' of the yarns – that is the increased length from one intersection to the next – gives an improvement in tear strength, but at the same time, the total number of warp/weft intersections is reduced. The number of intersections has a very significant effect on the ease with which an object piercing the fabric, such as a fastener, can comb threads out of the woven structure, giving lower fastener retention strength. This can be improved in the twill by breaking the regularity of the pattern, giving a 2/2 broken twill or 'crowfoot' weave. Here, the interlacing pattern of the third and fourth ends of the basic twill design is reversed.

Satin weave is achieved by 'floating' the warp or weft yarn over three or more of the opposite yarn. The floating yarn is then passed under one of the opposite yarns before repeating the process again. In the crowfoot satin weave, the weft yarn passes over three warp yarns and under one. It is more pliable than a plain weave and forms easily around curves.

Leno weave is a weave in which two warp yarns twist and grip tightly around the weft yarns. The major advantage of this weave is that the structure is very stable, with both warp and weft yarns held firmly, allowing much more open fabrics to be produced.

Conveyor belts may be classified into the following tension categories:

Low	0–105 kN/m
Medium	105–280 kN/m
High	280–560 kN/m
Ultra	560–1,050 kN/m

Woven fabrics such as plain, rib, basket and crowfoot are the most common styles in the low tension range of belts (0–105 kN/m). For medium tension range of belts (105–280 kN/m), uni-ply constructions such as straight warp and solid woven are used. Steel cord belts are designed for the medium- to ultra-tension ranges.

Earlier, cotton was the most popular material used for conveyor belts production. Afterwards, cotton was replaced by rayon. With the advent of synthetic reinforcement,

Table 4.1 Fabrics, their composition and characteristics

Name	Composition	Relative characteristics
Cotton	Natural cellulose	Low specific strength Increases in strength when wet Fairly high degree of elongation High absorption of moisture Consequently poor dimensional stability Poor rot, acid and weather resistance Limited impact resistance Susceptible to mildew attack Widely used in past
Rayon	Regenerated cellulose	Slightly stronger than cotton Similar to cotton in chemical resistance Moisture absorption is higher than cotton Tensile strength is lowered by water Susceptible to mildew attack Almost non-existent in conveyor belt today
Nylon (trade name: PERLON)	Polyamide (synthetic)	High strength and high elongation Good resistance to abrasion, fatigue and impact Low moisture absorption High resistance to mildew
Polyester (trade names: TERYLENE, DACRON, TREVIRA, etc.)	Polyester (synthetic)	High strength but not quite as high as nylon Good abrasion and fatigue resistance Extremely low moisture absorption Consequently good dimensional stability High resistance to mildew Generally acid/alkali resistant
Steel	Steel	Used where high strength and extremely low stretch are a necessity Small amount of woven steel carcass is in use Excellent heat resistance. Good fatigue and abrasion resistance. Superior troughing characteristics.
Kevlar	Aramid	Has twice the strength of steel Stretch characteristics roughly halfway between steel and polyester Tenacity is seven times as steel tenacity Lower in weight than steel and will not rust High abrasion resistance
Glass	Glass	Very high strength compared to rayon Low elongation Mainly used in high-temperature applications Poor flex life Limited use in belt manufacture currently
Asbestos	Asbestos	Low in strength Used for high-temperature applications

conveyor belt performance has been enhanced several times over. Major fibres currently used are cotton, rayon, nylon, polyester, steel, Kevlar, glass and asbestos. The composition and relative characteristics of these fibres are given in Table 4.1.

The fabric carcass may be made from one type of fibre or may use combinations as per the required characteristics. The following are the three important features of any conveyor belt carcass:

1. The elastic properties of the carcass when tensioned
2. The additional stresses of the carcass due to geometrical conditions of the conveyor
3. The strain of the carcass developed by the material conveyed

Due to the above factors, the conveyor carcass will stretch during operation because of changes in the belt tensions caused by starting and/or braking conditions or changes in thermal conditions. This is termed the elongation of belt. The per cent elongation at break will be many times more than the elongation at rated tension. Elongation of carcass is different from elongation of cover rubber. The elongation at break of cover rubber may be as high as 400%–500%, whereas carcass elongation of textile belt at rated tension would be in the range of 1%–4%. A conveyor belt can have the following types of stretch or elongation:

1. **Elastic Elongation**: This elongation is that part of stretch which occurs in a conveyor belt during starting acceleration, braking deceleration and belt loading. This elongation is almost entirely recovered when the applied pull or stress is removed.
2. **Constructional Elongation**: This elongation is more due to the type of fabric weave than to the textile material used. In a conventionally woven fabric, the warp strands that are crimped tend to straighten out as the load is applied. This results in belt growth, a portion of which is non-recoverable.
3. **Permanent Elongation**: This elongation includes changes in length caused by elongation in the basic fibre structure. It also includes that portion of the elastic elongation and constructional elongation which is non-recoverable.

Take-up systems allow the conveyor to compensate for these changes in overall belt length without having to cut sections out of the belt.

4.3 Types of conveyor belts

Conveyor belts can be classified principally into two types based on the carcass used as:

- **Fabric/Textile Belts**, consisting of layers of plies of fabric–duck, impregnated with rubber and protected by rubber cover, on both sides and edges. The fabric–duck supplies the strength to withstand the tension created in carrying the load, while the cover rubber protects the fabric carcass.
- **Steel Cord Belts**, made with a layer of parallel steel cords completely embedded in rubber. They can withstand much higher tensions and give a considerably longer life.

4.3.1 Fabric/Textile belts

The fabric/textile belts can be one-ply and multi-ply. In a one-ply belt, fabric ply is at the centre in between the top and bottom cover rubber and is shown in Figure 4.4. In a typical belt, a thin adhesive rubber layer (also called skim) is also applied at the bottom and top of the ply (Figure 4.5) so that rubber spreads very effectively through interstices of the fabric. Ply with adhesive rubber layers is kept in between the top and bottom covers and subjected to the hot vulcanization process which creates bond between ply and covers and results a belt. The adhesive rubber layer is an important contributor to internal belt adhesions and impact resistance and plays a significant role in determining belt load support and troughability. Improper or marginal adhesive rubber layer can adversely affect the belt performance in general and can lead to ply separation and/or failure at idler junction.

Breaker is the additional transverse layer of open-weave fabric used primarily to increase the adhesion between the cover and the carcass. Breaker fabric tends to dissipate impact energy and help to prevent puncture of the belt carcass from sharp material through the cover. It also prevents the progress of rubber cut. Placement of the breakers is generally between the carcass and the top cover for heavy abuse constructions helping to absorb the loading impact. Breakers can be made from the same cotton or synthetic fibres used in carcass fabrics. Figure 4.6 shows single-ply textile belt with breaker.

Because cover adhesion increases due to addition of breaker, sometimes it is provided in the bottom cover also for improving its adhesion to the carcass. This improves the belt life when high tractive pull is to be transmitted from drive pulley to the carcass. The use of modern synthetic materials such as polyester and nylon has essentially eliminated the need for the breaker strip.

In multi-ply belts, an adhesive rubber layer is applied between plies. This layer between plies improves the ability of the belt to withstand impact by diffusing the impact forces. This results in more durable bond between plies and is used for handling heavy/lumpy materials. Figure 4.7 shows two-ply and three-ply belts with inter adhesive rubber layers.

One-ply belts of general construction are used for certain low-tension applications. Two-ply belts are used for medium belt tensions. Multi-ply belts are used for heavy-duty conveyors for the machines such as bucket wheel excavators, stackers and reclaimers.

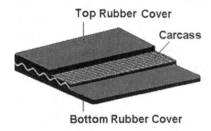

Figure 4.4 One-ply belt.

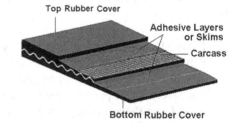

Figure 4.5 One-ply belt with adhesive layers.

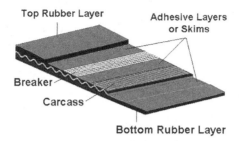

Figure 4.6 One-ply belt with breaker and adhesive layers.

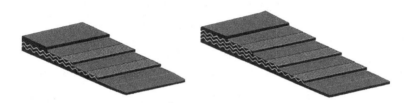

Figure 4.7 Two-ply belt and three-ply belt with adhesive layers.

There are two methods of manufacturing fabric conveyor belt, i.e. **moulded edge** and **cut edge** constructions (Figure 4.8). In the moulded edge construction, the carcass terminates slightly before the belt edge and the belt edges are moulded/made from rubber. Closing of the belt edge is necessary to prevent ingress of moisture. This construction was very common during the use of cotton fabric as carcass material because cotton fabric is susceptible to moisture spread by capillary with consequent weakening due to biotic effect. Moulded edge is also necessary if the carcass is to be shielded from the environmental condition. However, these belts are costly because moulding of the edge involves additional step in manufacture. While using synthetic carcass, it is not necessary to close the belt edge as even with open edge, moisture is unlikely to be

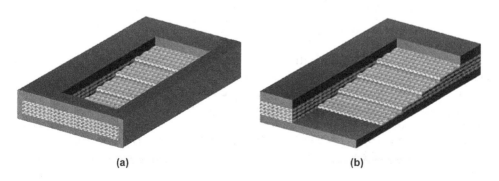

(a)   (b)

Figure 4.8 (a) Moulded edge and (b) cut edge construction of belt.

absorbed. This open edge is called cut edge. A cut edge belt is manufactured and then cut down to the specified width. As a result, cut edge belt is cost-effective and becomes more common. The drawback of this cut edge belt is that the carcass of the belt is exposed; therefore, the carcass is more vulnerable to the problems arising from abusive environmental conditions during storage, handling and use.

Carcass is made with cotton, nylon and polyester and combinations of these. Some of the popular versions are as follows:

a.   Cotton fabric (warp cotton, weft cotton)
b.   Cotton–nylon (warp cotton, weft nylon)
c.   Nylon–nylon (warp nylon, weft nylon)
d.   Polyester–nylon (warp polyester, weft nylon)

Cotton fabric belts continue in usage in spite of their following disadvantages:

•   Low strength-to-weight ratio
•   Low lateral stiffness, which limits the troughability angle
•   Lower longitudinal troughability, which is characterized by the usage of larger pulley diameters
•   Only mechanical adhesion between the cotton carcass or plies of the belt and the rubber, resulting in lower adhesion
•   Cotton fabrics having the tendency to rot in the presence of water, dirt, etc.

To overcome the obvious disadvantages of cotton, cotton–nylon fabrics were developed. These employ nylon in the core as the reinforcement embedded in a matrix of cotton. The advantages are improvement in flexibility in longitudinal and transverse direction resulting in lower pulley diameters. However, these belts are continued to suffer from the disadvantages of low strength-to-weight ratio and low adhesion with rubber as adhesion was achieved again by mechanical linkage between the cotton and rubber. Also such belts have a tendency to rot when the carcass was exposed.

Nylon–Nylon belts (NN belts) have nylon fabric as the warp as well as weft members. The advantages of these belts are as follows:

•   Excellent strength-to-weight ratio
•   Good impact and flex fatigue resistance
•   Hardly any effect by moisture and mildew
•   High adhesion between fabric and rubber

Though nylon has a higher elongation compared to cotton, this could be reduced to an acceptable level by heat setting under tension.

Due to the higher elongation of nylon, fabric with polyester as warp member and nylon as weft member (called PN belts) were developed to produce belts with lower elongation, particularly for short length conveyors, where take-up facility is minimum. As polyester comes from ethylene glycol (E) and nylon from polyamide (P), these belts are also called as EP belts. In addition to lower elongation, other advantages are as follows:

- Greater resistance to the effects of moisture and water
- Excellent toughness and high strength
- Excellent resistance to impact
- Excellent troughability and load support
- Excellent resistance to chemicals
- Excellent fastener holding ability and flex life

There are fabric belts of numerous designs. However, the major types of belt carcasses are multi-ply belt, reduced-ply belt and solid-woven belt.

1. **Multi-Ply Belt**: Multi-ply belt is the most popular belt construction and is usually made up of two or more plies of woven cotton, rayon or a combination of these fabrics bonded together by an elastomer compound. Belt strength and load support characteristics vary with the number of plies and the fabric used. The multi-ply belt was the most widely used belt through the mid-1960s, but today it has been supplanted by reduced-ply belt.

2. **Reduced-Ply Belt**: Reduced-ply belt consists of carcass with either fewer plies or special weaves. The fabric carcasses of such belts comprise high-strength synthetic fibres, usually nylon, polyester or combinations, in plain weave or special fabric-designs. Load support is provided by extra thick rubber skim coat between plies and/or in the weave design of the fabric itself. The technical data of belt manufacturers generally indicate that reduced-ply belt can be used for the full range of applications specified for multiple-ply belt.

3. **Solid-Woven Belt**: Solid-woven belt consists of a single ply of solid-woven fabric impregnated and covered with PVC with relatively thin top and bottom covers (Figure 4.9). Abrasion resistance of PVC is lower than that of rubber. Hence, some solid-woven belts are made with a combination of a PVC core and rubber covers. These belts are primarily used in underground conveying, especially as impregnated by PVC for flame resistance applications.

Figure 4.9 Solid-woven belt with PVC cover. (Courtesy – Fenner Dunlop Engineered Conveyor Solutions.)

Conventional plied belts employing all synthetic carcasses and elastomeric covers are particularly recommended for hard rock mining, aggregate, sand, ore, general-purpose applications, forest products, soft minerals such as coal, potash, phosphates and grain.

4.3.2 Steel cord belts

Steel cord conveyor belts are made with a single layer of parallel steel cords (also called as steel wire ropes or steel cables), completely embedded in bonding rubber matrix, as the tension element. There are mainly two different types of cords. One is a 7 × 7 cord for low-to-mid tension, and the other is a 7 × 19 cord for mid-to-high tension. These two types of cord enable rubber to fully penetrate to the centre core. If rubber does not penetrate to the core, the cord may corrode from small cuts or at the belt ends. The steel wires are usually made of high-carbon steel and have a surface finish, which facilitates adhesion to the surrounding rubber and reduces corrosion during use.

In the 7 × 7 cord, 7 steel wires are twisted together to form one strand. Then, 7 such strands are twisted together to form the cord. In the 7 × 19 cord, 19 steel wires are twisted together to form one strand. Then, 7 such strands are twisted together to form the cord (Figure 4.10). In a cord, the direction of twist for the strand is opposite to the direction of twist for the cord. Such cords are utilized to make a steel cord belt. In a steel cord belt, the steel cords are spaced parallel in a single layer and the cords in a belt are arranged with opposite twists side by side as shown in Figure 4.11. This arrangement eliminates accumulation of residual torsion and also minimizes possibility of belt mistracking. The steel cord belts are made with moulded edge construction.

Steel cords are used in belts when textile fibres in cord or fabric form are unable to satisfy the engineered design or the service conditions. The steel cable wires are normally made from high-carbon steel. To obtain a better bond between the wire and the surrounding rubber stock, a special finish is used on the wire.

Formerly, steel cords or small steel ropes were used to reinforce high-tension textile belts. However, not only can steel cords withstand much higher tensions than fabric plies, but also the actual design of a modern-day fabric-free steel cord belt gives a considerably longer life, especially when heavy or sharp-edged material is to be conveyed.

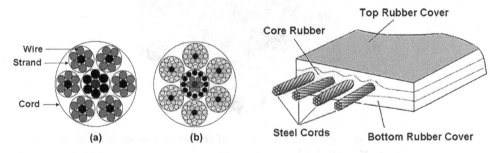

Figure 4.10 Steel cords. (a) 7 × 7 cord and (b) 7 × 19 cord.

Figure 4.11 Cross section of steel cord belt.

Figure 4.12 Troughing ability of steel cord belt.

The steel cord belt allows very thick covers and is still very flexible. Because of the all-rubber construction and absence of weft members, steel cord belt troughs very easily (Figure 4.12).

However, steel cord belts in some cases need additional protection. Additional means in addition to its own protection effected by the all-rubber design. Sometimes rods or other metal parts may penetrate a steel cord by force and, getting caught underneath, may slit the belt lengthwise. Though repair is simple in such cases, it means a lot of downtime. Consequently, a number of crosswise reinforcements are applied for steel cord belts as follows:

1. Normal steel cord belt, rubber and cords only (Figure 4.11)
2. Steel cord belt with envelope of fabric–duck (Figure 4.13)
3. Steel cord belt with diagonal layers of textile cord (Figure 4.14)
4. Steel cord belt with crosswise layer of wire strand at the top (Figure 4.15)
5. Steel cord belt with crosswise layers of wire strand at the top and at the bottom (Figure 4.16)
6. Steel cord belt with crosswise layer of high-tensile textile yarn at the top (Figure 4.17)
7. Steel cord belt with crosswise layers of high-tensile textile yarn at the top and at the bottom (Figure 4.18)

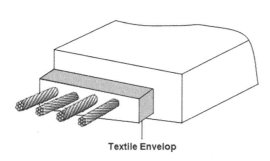

Textile Envelop

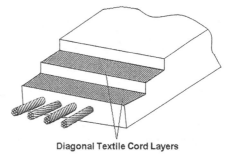

Diagonal Textile Cord Layers

Figure 4.13 Steel cord belt with envelope of fabric–duck.

Figure 4.14 Steel cord belt with diagonal layers of textile cord.

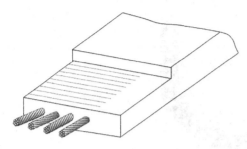

Figure 4.15 Steel cord belt with crosswise layers of wire strand at the top.

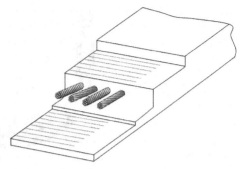

Figure 4.16 Steel cord belt with crosswise layers of wire strand at the top and at the bottom.

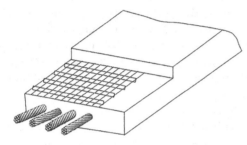

Figure 4.17 Steel cord belt with crosswise layers of high-tensile textile yarn at the top.

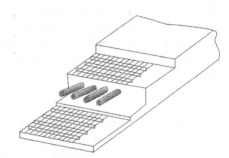

Figure 4.18 Steel cord belt with crosswise layers of high-tensile textile yarn at the top and at the bottom.

Among the above, steel cord belts with crosswise layers of highly tensile textile yarn (type 6 and 7) are the optimum.

4.3.2.1 Special features of steel cord belts

The following are some of the special features of steel cord belts:

1. High working tension
2. Long span
3. Good troughability
4. Low elongation
5. Good dynamic performance and long belt life
6. Longer life of splicing
7. Easy internal inspection of belt and splice
8. Reconditionability
9. Smaller pulley diameters
10. High impact resistance

1. **High Working Tension**: In comparison with the availability of maximum working tension of 900 kg/cm width with steel cord belt, the maximum working tension of nylon belt, which is used nowadays on a large scale, is about 300 kg/cm width. Thus, steel cord belt has three times the working tension of the textile carcass belt.

2. **Long Span**: It is possible to have longer belts with steel cord belts. The longest possible conveyor lengths of different belts are given below for comparison:

Steel cord belt	25,000 m
Nylon belt	4,500 m
Cotton belt	600 m

Longer lengths mean less of drive heads, loading points, etc., which is not only economical but also minimizes the problems.

3. **Good Troughability**: Steel cord belt has very good troughability compared to fabric belts. In a fabric belt, one of the major damages is the edge damage due to crooked running of the belt because of poor troughability. Recently, thin and strong fabric belts, either nylon or polyester belts, are used for general purpose, and their troughability has been improved considerably.

4. **Low Elongation**: This is one of the most outstanding features of steel cord conveyor belt. The steel cord conveyor belt is most useful where the installation of the conveyor system requires an extremely limited take-up travel. In comparison, steel cord belt has very little elongation of about 0.1–0.2% and the take-up movement can be very small, as compared to fabric conveyor belt. Surging is also limited in steel cord belts.

5. **Good Dynamic Performance and Long Belt Life**: Steel cord belt has no ply-type tension members such as fabric belt. The adhesion between steel cords and rubber is quite strong due to deep permeation of rubber. Hence, the steel cord is superior in flexural fatigue resistance and shock resistance than fabric belts.

6. **Longer Life of Splicing**: A remarkable characteristic of steel cord belt is that the life of spliced part is long, and hence, the total life of the belt as a whole is much higher than fabric belts. In the case of fabric belt, the over-ply is tensioned, while the inner ply is compressed. Hence, repeated flexing around pulleys, with repeated shocks, the rubber between plies, particularly in the splicing area is likely to separate. Steel cord belts, however, having only a single layer, are almost free of such fatigue.

7. **Internal Inspection of Belt and Splice**: In a textile belt, it is difficult to observe the inside condition of the belt or the splice, whereas in the steel cord belt, a non-destructive testing method such as sending X-rays through the belt can be employed to examine the arrangement of steel cords, damage to the steel cords, displacement of steel cords and slight separation of steel cords from rubber, etc., without cutting the belt.

8. **Reconditionability**: Both textile and steel cord belts can be reconditioned economically to achieve substantial working life, if the belts are removed at the correct time for reconditioning.

9. **Smaller Pulley Diameters**: Pulleys of smaller diameter can be used for operating steel cord belts as they can easily flex over the pulley.

10. **High Impact Resistance**: As the carcass is made of steel cords, obviously steel card belts have high impact resistance.

4.3.3 Metaflex belts

These are the belts with single layer of steel mesh or mat as the carcass (Figure 4.19). These belts have been developed for critical operating conditions such as sinter plant and coke oven applications.

The advantages of these belts are as follows:

1. High impact resistance
2. High heat/burn through resistance
3. High adhesion and consequent superior belt consolidation
4. Low self-weight
5. Improved transverse flexibility
6. Higher longitudinal flexibility
7. Superior joint efficiency
8. Low belt stretch

4.3.4 KEVLAR belts

For conveyor belts, a new reinforcing fibre, aramid (aromatic polyamide), called by the brand name of KEVLAR, has been introduced. KEVLAR aramid fibres were developed by Du Pont for use in application where high strength, high modulus, low elongation in use and light weight are important belt selection criteria. On a weight basis, KEVLAR is five times as strong as steel.

These properties, combined with the fibre's resistance to corrosion, inherent flame resistance, low thermal conductivity plus good fatigue and chemical resistance, enable KEVLAR to be used as reinforcing fibre in conveyor belts. Figure 4.20 shows a KEVLAR belt with breaker.

KEVLAR aramid fibres have been tried to construct the carcass of plain woven, woven cord, straight warp weave, solid woven and cabled cords, and the best results were obtained for the latter four constructions.

Advantages of KEVLAR belts over steel cord belts

1. For the same weight, KEVLAR is five times stronger than steel
2. KEVLAR does not corrode in wet environments
3. KEVLAR has no sparking potential and hence is safer for use in underground coal mines where flammable gases may be present
4. KEVLAR aramid fibres are able to absorb the impact better when the belt is charged, thereby reducing wear on the rubber covers

Advantages of KEVLAR belts over textile belts

1. Higher strength
2. Lighter weight
3. Improved non-flammability
4. Improved safety

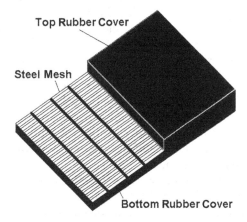

Figure 4.19 Metaflex belt.

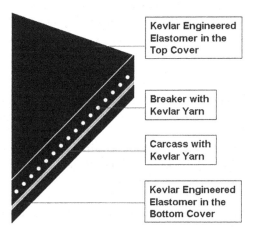

Figure 4.20 KEVLAR belt.

4.4 Top and bottom covers of the belt

By incorporating rubber layer on top and bottom of the carcass, carcass is protected from material impact, wear, abrasion and environmental effects due to moisture, sunlight, ozone, chemicals, heat, cold and petroleum products, etc., so that it does not weaken and extend its service life. The rubber layer on the material-carrying side is known as **top cover**. The rubber layer on the bottom side (non-carrying side or running side) is known as **bottom cover**. The top and bottom covers of the conveyor belt provide very little, if any, structural strength to the belt. The cover is usually the lowest cost component of the belt. The main purposes of top and bottom covers are as follows:

1. to protect the carcass from impact damage

   When the material falls onto the belt, when loaded, its sharp edges tend to cut/gouge the belt. The top cover protects the carcass by withstanding such impact forces.

2. to protect the carcass from wear

   a.   After the material touches the belt, when loaded, it rubs with the belt for short time till the material velocity matches the belt velocity. The cover bears this abrasive action.

   b.   There is a minute to-and-fro movement of the material when the belt passes over the idlers during running of the belt. The top cover withstands the wear due to this material movement.

   c.   There is a slight but continuous wearing effect on the belt as it moves on the idlers on both the carrying and return sides. This effect is taken by the bottom cover during carrying run and by the top cover during return run.

   d.   The wearing effect also arises due to material contamination, stuck and misalignment of idlers.

3. to provide a friction surface for driving the belt
4. to protect the belt against any special deteriorating factors that may be present in the operating environment

Rubber or rubber-like compounds are used for the top and bottom covers of conveyor belt and for bonding together various components of the belt carcass. These compounds are produced by mixing rubbers or elastomers with various chemicals in order to obtain reinforcement and to develop the physical properties necessary for service conditions.

Materials that are able to return to their original shape after deformation with a minimal permanent distortion are known as **elastomers**. In the case of conveyor belt, this is extended to include all the thermosetting materials such as natural rubber, synthetic rubbers and thermoplastic materials such as PVC (polyvinyl chloride) and plastic, which require definite time and temperature for cure. A few elastomers and their characteristics are given in Table 4.2.

It is possible to classify covers by the basic elastomer used, but the evaluation of the quality of the cover should be based on its suitability for particular service, rather than on the kind of elastomer it contains. Each cover has characteristics which, when properly utilized, will provide a conveyor belt for the lowest cost per unit of material carried under specified conditions of service.

The factors that must be taken into account when selecting the belt grade or cover rubber are as follows:

1. Fire resistance or anti-static properties
2. Resistance to oils or chemicals
3. Resistance to ageing, weathering and ozone
4. Temperature of the operating environment or conveyed material
5. The type of material conveyed
6. The lump size and shape of the material conveyed
7. The mix of lumps and fines in the material

Table 4.2 Elastomers and their characteristics

Elastomer	Characteristics
Natural rubber (NR)	Very good tensile strength and elongation High heat resistance and elasticity High tearing and shear strength Good abrasion resistance Stable within the temperature range of −30°C to +80°C. Resistant to water, alcohol, acetone, dilute acids and alkalis
Synthetic rubber, styrene–butadiene rubber (SBR)	Tensile strength and cut resistance are good. Abrasion, heat and ozone resistances are better than natural rubber.
Nitrile rubber, nitrile–butadiene rubber (NBR)	Relatively abrasion resistant, resistant to ageing
Butyl rubber, copolymer of isobutylene and isoprene rubber (IIR)	It has a very good ozone and temperature resistance Able to withstand temperatures of −30°C to +150°C Limited resistance to acids, alkalis and animal and vegetable fats
Ethylene propylene rubber (EPM) or ethylene propylene diene rubber (EPDM)	Temperature resistant Higher resistance to wear and tear Better ozone resistant than all other basic polymers
Chloroprene rubber (CR) commonly called as neoprene	The mechanical properties are similar to natural rubber but significantly better in respect of ozone and oil resistance. The chlorine in chloroprene gives the product a high degree of flame resistance. Resistance to animal and vegetable oils and fats is superior to natural rubber.
Nitrile–chloroprene rubber (NCR)	High oil and fat resistance Superior mechanical properties
Silicone rubber (VMQ), fluorosilicone rubber (FVMQ)	Outstanding resistance to high heat Excellent flexibility at low temperatures Very good electrical insulation Excellent resistance to weather, ozone, sunlight and oxidation Poor resistance to abrasion, tear and cut growth Low tensile strength Inferior resistance to oil, gasoline and solvents Poor resistance to alkalis and acids
Chlorosulfonyl polyethylene (CSM)	Excellent ozone, weathering and acid resistance Good chemical resistance Good abrasion, cold, heat and flame resistance

8. The abrasiveness of the material
9. The method of feeding the belt
10. The fall height of material to the belt
11. The cycle time of the conveyor for a single revolution of the belt
12. Availability and cost

4.4.1 Top and bottom cover quality

There are two standard grades of belt covers as well as special grades designed for difficult circumstances. The standard grades of cover compounds are defined by Grade N and Grade M as per Indian Standard and RMA Grade I and RMA Grade II as per The Rubber Manufacturers Association of the USA and belting industry. The widely used cover grades are **Grade M** and **RMA Grade I** and **Grade N** and **RMA Grade II.**

Grade M and RMA Grade I cover rubbers consist of natural or synthetic rubber or blends and are characterized by high cut, gouge and tear resistance and very good-to-excellent abrasion resistance. Grade I cover conveyor belt has a skim coat of rubber compound between the plies. The type of rubber compound used with the particular fabric insures the highest degree of flex life. These covers are recommended for service involving sharp and abrasive materials, and for severe impact loading conditions.

Grade N and RMA Grade II cover rubbers consist of elastomeric composition similar to that of Grade M and RMA Grade I with good-to-excellent abrasion resistance but not as high a degree of cut and gouge resistance as that of Grade M and RMA Grade I covers. Grade II cover conveyor belt also has a skim coat of rubber compound between the plies. This is ideal for general class of materials, which are not very lumpy or sharp-edged.

The wear resistance quality of a conveyor belt is one of the major factors that determine its life expectancy. As a general rule, 80% of conveyor belt surface wear usually occurs on the top cover of the belt with approximately 20% of wear occurring on the bottom cover. Wear on the top cover is primarily caused by the abrasive action of the materials carried, especially at the feeding point where the belt is exposed to impact by the bulk material and the material is effectively accelerated by the belt surface.

Belt cleaning systems such as scrapers can also cause wear to the top cover surface. Wear on the bottom cover is mainly caused by the friction contact made with the drum surface and idlers. Factors such as unclean environment, where there is a build-up of waste material, can cause added wear on both the top and bottom covers of a belt. Short belts usually wear at a faster rate than long belts due to the higher loading frequency. This is because the belt passes the loading point more frequently compared to longer length belts.

Factors that influence the cover wear are as follows:

1. The distance from which the material falls onto the belt at the feeding point, and the design of the feeding point itself
2. Increased abrasion occurs, if the material is transferred at right angles to the belt travel
3. Increased abrasion occurs, if the material is fed on to the belt where the incline is greater than 8°
4. Wear is increased if the material is sized or screened, or largely contains lumps with few fines
5. The passing of the belt over return idlers, which have become roughened or corroded, or have seized due to lack of maintenance
6. The shifting of the load as it passes over the carrying idler. This factor can be accentuated by insufficient tension to limit sag between idlers

The two other most important characteristics are resistance to tear and resistance to abrasion. Tear strength is important where sharp objects such as rocks, metals and timber are transported, which can cause cutting and gouging. Abrasion resistance is especially relevant where fine materials such as aggregate, sand and gravel are carried, because these materials tend to create a highly abrasive effect in a similar way to sandpaper when they fall onto the belt.

Heat-resistant belts are designed to carry materials at high temperatures with minimum of ageing. The cover compounds of heat-resistant belts consist of butyl or ethylene propylene (EPM) rubber, which can resist degrading effects of high temperatures up to 400°F. Neoprene (polychloroprene)- and hypalon (chloro-sulphonated polyethylene)-based compounds also exhibit good heat ageing properties. Belt with silicone or Viton (fluorocarbon polymers) covers will withstand very high temperatures, up to approximately 700°F.

Fire-resistant belts are constructed using materials that do not continue to burn once they have been ignited and the source of the flame has been removed. SBR, nitrile, polychloroprene (neoprene) and PVC are routinely utilized to make fire-resistant belts.

Conveying materials that contain oil and fat can have a very detrimental effect on the performance and life expectancy of a conveyor belt because it penetrates into the rubber causing it to swell and distort often resulting serious operational problems. Belt covers designed to resist swelling and degradation in oily environments will often incorporate nitrile-based polymer, polyvinyl chloride (PVC) or urethane. These covers are also of hygienic grade and used to handle foodstuffs.

Fire-resistant- and oil-resistant-grade rubbers often have lower resistance to wear and impact, and therefore, these should be used where needed. Generally, they wear faster and also cost more compared to usual grade – M or N.

4.4.2 Thickness of the belt covers

The cover of a conveyor belt must be sufficiently thick to withstand gradual wear even for long period and to protect the tension member from mechanical injury. The top cover will generally be greater in thickness than the bottom cover because operational strain and wear are much more in the top cover. However, specific characteristics of the material to be conveyed and the operating conditions may require a belt with equal cover thickness on top and bottom.

There is a general tendency to give less attention to the bottom cover. In fact, it is most important to have adequate thickness for the bottom cover, taking into consideration the pulley and return slippage, accidental tears and bites and operation under unfavourable conditions. Generally, the thickness of the bottom cover should be between 40% and 60% of the thickness of the top cover. In the case of steel cord conveyor belts, the thickness of the bottom cover must be 4mm minimum, and the standard thickness is 60%–100% of the top cover.

Cover thickness is to be increased with an increase in abrasiveness, maximum lump size and weight of the material, height of material drop onto the belt, feeding angle, belt speed and frequency of feeding as determined by frequency factor. The frequency factor is the number of minutes for the belt to make one complete turn and is determined as

$$F_f = \frac{2L}{V}$$

where F_f is the frequency factor in minutes, L is the centre-to-centre length of the belt conveyor in feet and V is the belt speed in feet per minute.

The thickness of the covers for special type of conveyor belts, such as heat-resistant, oil-resistant, rough top belts, requires special consideration.

4.4.3 Cover quality and thickness for replacement belts

If the belt under consideration is replacement for an existing one, analysis of the previous causes for failure will help to determine the thickness and grade of the new belt, for example:

1. If the carcass is sound but the top cover is worn by abrasion, a thicker or better quality cover is required
2. If the cover failed owing to separation from carcass, this suggests that the cover thickness is adequate, but that a tie cloth (breaker) is required to increase the adhesion between cover and carcass
3. If the carcass is damaged through impact at the loading point, a thicker cover in conjunction with a breaker should be used and/or heavier carcass is required

4.4.4 Types of belts and their applications

The following are the few types of the belts normally in use:

1. **Standard Rubber Belts**: Standard belts incorporate covers suitable for the handling of most abrasive materials having a blend of natural and synthetic rubber.
2. **Cut-Resistant Belts**: Cut-resistant belts have a high content of natural rubber, recommended for belts operating under extremely difficult conditions where cutting and gouging of covers may occur.
3. **Heat-Resistant Belts**: These belts incorporate covers with styrene–butadiene, recommended for belt handling materials with temperatures up to 1,200°C.
4. **Super-Heat-Resistant Belts**: These belts have chlorobutyl covers, recommended for belt handling materials with temperatures of up to 1,700°C.
5. **Fire-Resistant Belts**: Fire-resistant belts are manufactured with covers containing neoprene with multi-ply carcass constructions to meet the maximum standards of safety in underground mines.
6. **Wood Handling Belts**: These belts were especially developed non-staining for the timber industry compounded to provide resistance to oil and resin.
7. **PVC Solid-Woven Belts**: PVC solid-woven belts are manufactured with polyester and nylon with cotton soaked in armour of PVC and PVC coatings, developed to withstand the impact, tear and abrasion as the requirements to meet the most stringent flame-resistant standards.
8. **Food-Quality Belts**: Food-quality belts are manufactured from non-toxic materials resistant to oils, fats and staining to meet the hygiene requirements of the food processing industry.

9. **Nitrile-Covered PVC Belts**: Nitrile-covered PVC belts were developed for application in mines where the danger of fire exists and also have properties of flame retardant, oil, abrasion and heat resistance.
10. **Steel Cord Belts**: Steel cord belts are generally manufactured for application in long-distance conveyors stiffened with a steel wire inserted within a high-quality rubber in order to get exceptional traction load and material high impact.
11. **Fire-Resistant Steel Cord Belts**: Fire-resistant steel cord belts were developed with properties of self-extinguishing fire to offer advantages in free maintenance operations and long life for conveyors situated in fiery mines.
12. **Oil-Resistant Belts**: These belts are manufactured to provide easily washable linings in nitrile, neoprene or synthetic rubber in all layers, allowing ease of application in handling of materials containing vegetable oils and minerals.
13. **Chemical-Resistant Belts**: These belts are used for transporting chemicals, pulp, ceramic, foodstuffs, fertilizer and materials with chemicals attached. It is necessary to select cover rubber that is resistant to acid or alkali depending on the types of transport materials or chemicals attached to the materials.
14. **Anti-Static Conveyor Belts**: These belts are made of cover rubber especially mixed to prevent static electricity. Anti-static belt is essential to transport fabrics that stick on the belt surface with static electricity or electronic products that may explode or ignite by electrification.

4.5 Factors to be considered for the belt design

4.5.1 Tension rating

The tension rating for a belt is the recommended maximum safe working stress that should be applied to the belt. Conveyor belts are designed to withstand the operating tensions, and the belts are rated accordingly. These ratings are standardized and maximum operating tensions for various belts are tabulated and supplied by the belt manufacturers.

4.5.2 Load support

Full load of material fed on to the conveyor belt must be supported by the belt as the belt spans between two idler sets. Belt manufacturers supply tables reflecting suggested belts for various types and grades of material to be transported on the belt.

4.5.3 Troughability

The belts are troughed to allow the conveyor to carry more material. The ability of the loaded belt to trough properly on the carrying idlers is called troughability. Conveyor belt should not be too stiff or too thin. Transverse flexibility or rigidity of the belt is important for the belt to trough properly. The empty conveyor belt must make sufficient contact with the centre roller in order to track properly. As shown in Figure 4.21a, the top belt is too stiff to contact the centre rollers and therefore will wander from side to side with the possibility of causing considerable damage to the belt edges and to the structure. The belt shown in Figure 4.21b shows sufficient contact with the centre roller and is the required condition. As a general rule, the belt troughability is adequate if minimum 35%–40% of the belt width is in touch with idlers, even when the belt is empty. In this condition, it should also touch the central roller. Transverse flexibility varies with the belt width and trough angle. The belt with stronger fabric and lower number of plies is good for flexibility. Troughability also increases with an increase in belt width.

Most conveyor belts operate over troughed idlers. The troughing angle of these idlers varies from 20° to 45° and beyond as shown in Figure 4.21c and d. The trough angle affects the belt by creating a line along which the belt is constantly flexed. The greater the trough angle, the greater the flexing action. When the belt is fully loaded, the portion of the load over the idler junction gap forces the belt to flex to a shorter radius. The heavier the load, the smaller the radius through which the belt must flex. At higher troughing angles, gravitational force is exerted on the portion of the belt in contact with the wing roller. All these forces try to push the belt down into the idler junction gap. As shown in Figure 4.21e, the belt design is satisfactory in that it does bridge the angle properly under full load, whereas as shown in Figure 4.21f, the weight of the load has forced the belt tightly into the gap between rollers and premature failure can occur. Hence, a belt is to be designed with sufficient transverse rigidity to bridge the idler junction gap with a satisfactory radius and flex life so that for a given idler angle and load weight, premature belt failure will not occur.

4.5.4 Quality and thickness of the belt covers

Belt covers must consist of the correct grade of material and be of sufficient thickness to suit the application. Belt manufacturers supply tables reflecting suggested minimum top cover thickness based on an impact exposure frequency factor as well as

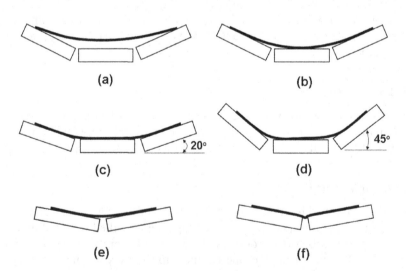

Figure 4.21 Conveyor belt troughability analysis. (a) Stiff belt, improper troughing; (b) flexible belt, proper troughing; (c) roller troughing angle 20°; (d) roller troughing angle 45°; (e) belt design unaffected by load weight; and (f) belt design affected by load weight.

characteristics of materials being conveyed. Bottom covers are not normally subjected to loading conditions, and in general, the minimum cover thickness as recommended by the manufacturer may be applied.

4.5.5 Belt width

For a given speed, belt conveyor capacity increases with the belt width. However, the width of the narrower belts is dictated by the size of lumps to be handled. Belts must have enough width so that any combinations of lumps and finer material will not load the lumps too close to the edge of the conveyor belt.

Belt width of the conveyor belt ranges from 300 to 3,600 mm. Certain minimum belt width is necessary in relation to lump size. If the belt width is not adequate, it will affect the belt and the idler life. It can also result in jamming at loading point.

The recommended maximum lump size for various belt widths is as follows:

	Maximum lump size
For 20° surcharge	
with 10% lumps and 90% fines	One-third of belt width
with all lumps, no fines	One-fifth of belt width
For 30° surcharge	
with 10% lumps and 90% fines	One-sixth of belt width
with all lumps, no fines	One-tenth of belt width

4.5.6 Belt thickness

Belt thickness is the sum of both the top and bottom covers and the thickness of the carcass. Belt thickness depends on various factors and would control the total life of the belt to a great extent.

4.5.7 Belt speeds

Belt speeds depend upon the characteristics of the conveyed material, the capacity desired and the belt tensions employed.

Relevant material characteristics include the abrasiveness of the bulk solid, its lump size and its tendency to dusting. At high belt speeds, general wear on the moving parts of the conveyor is greater, and especially with narrow belts, satisfactory belt tracking becomes increasingly difficult to maintain. Also, there is more risk of lumps rolling off the belt if it is running fast. Heavy sharp-edged materials should be carried at moderate speeds because the sharp edges are likely to wear the belt cover particularly if the loading velocity of the material in the direction of belt travel is appreciably lower than the belt speed.

Powdery material should be conveyed at lower speeds to minimize dusting, particularly at the feeding and discharge points. Similarly, fragile materials should also be conveyed at low speeds so that material degradation will not occur at the feeding and discharge points, as the conveyor belt and the material carried move over the rollers.

Typically, practical belt speeds are around 1.5 m/sec for very abrasive material or large lumps, up to 3 or 4 m/sec for free-flowing, non-abrasive products. Modern trends seem to be towards higher belt speeds because of the significant cost advantages that can be gained; 6 m/sec is fairly common, and up to 10 m/sec is possible in some situations. It is suggested that if care is taken over the dynamic design of the system, belt speeds above 15 m/sec are technically feasible.

4.5.8 Belt conveyor capacity

Belt conveyor capacity increases as the belt width increases for a given speed of the belt. The capacity of a belt conveyor depends on the surcharge angle and on the inclination of the side rolls of three-roll troughing idlers. If the feed to the conveyor is uniform, the cross-sectional area of the material load on the conveyor belt is the determinant of the belt conveyor capacity.

4.5.9 Minimum pulley diameters

When the pulley diameter is too small, belt is forced to excessively flex around the pulley circumference causing the development of internal stresses which leads to belt damage. This may result in separation of plies, ply failure, cracking of the top cover of the belt, pulling out of mechanical splices, etc., and eventually, failure of the structure of the belt takes place. Belt manufacturers supply tables reflecting suggested minimum pulley diameters for different belts.

4.6 Storage and handling of conveyor belts

Of all components of the belt conveyor, the conveyor belt is the most vulnerable to environmental effects and other operational forces. For this reason alone, the belt replacement cycle is comparatively shorter. The belt cost forms nearly 25%–35% of the total installation cost. This lays down the emphasis on the need for implementing a strict code of practice in respect of belt storage and handling to affect major belt performance enhancement.

When storing a new conveyor belt, leave it hoisted or stands it upright, preferably on a dry surface. A wooden skid is best. Block it safely so that it cannot accidentally roll. A dry place out of direct sunlight is preferred for storage, and excessive temperature variations or extremes are to be avoided. Belts should ideally be stored indoors in a temperature range of 0°C–25°C. Belts should not be stored in excessively wet places or in areas where oils, gasoline, paint, etc., are stored.

If belts are to be stored for long periods or in extreme climate, some protection from direct sunlight and extreme temperatures is recommended. At coastal sites, protection from ozone attack is also necessary. If proper warehousing is not possible, then belts should at least be covered with black polythene, suitably fastened against wind damage. The belt roll, if resting on the belt face, should be rolled a bit every six weeks to distribute the permanent deformation on the belt caused on the regions lying under the belt weight. Preferably, the belt rolls should be stored while mounted on A-frame after passing the 75-mm square centre bar. In general, it is wise to keep any unused belt stored in its protective factory packaging until it is ready for installation. Used belt should be thoroughly cleaned and dried prior to storage.

Conveyor belts are normally wound on wooden centred shell that accepts the 75-mm square bar for lifting. During transportation, the belt should be lifted on and off vehicles using a square centre bar. Wire ropes with hooks are applied to the bar on either side of the belt roll and a **spreader** above the roll to avoid damage to the belt which could otherwise take place due to wire rope cutting into belt edges starting the top of the belt roll. Belts should always be moved by lifting and transporting rather than rolling, due to the danger to men and the possibility of damage to the belt, and particularly, collapsing as a result of rolling. Should there be a reason still to roll the belt, it should strictly be done in the direction marked on the roll. Otherwise, the belt gets loosened in the roll. The belt can also be moved safely by laying the roll flat on a skid and hoisting the skid with a forklift. While doing so, be sure that the forks on the lift do not come in contact with the belt itself.

4.7 Belt splicing

The two ends of conveyor belt must be spliced (joined together) to form a continuous endless belt. When a belt is damaged or worn out, it needs replacement. As the conveyor length increases or decreases, the belt has to be extended or shortened and may need to add or remove one or more pieces of the belt. In such cases also, the belt ends have to be joined.

The durability of a conveyor belt largely depends on the splices made in the belt. These splices are in most cases weaker than the actual belt and are frequently the reason for conveyor disruptions. The strength of a splice depends on numerous factors including the following:

- The linkage layer
- The splicing method
- The materials used to make the splice
- The workmanship
- The environment in which the splice is made

Three common methods of splicing the belt are as follows:

1. Mechanical splicing where belt ends are joined by mechanical fasteners
2. Hot vulcanization where belt ends are together heated and cured under pressure with a vulcanizing press
3. Cold vulcanization, which uses a bonding agent that causes a chemical reaction to splice the two belt ends together

4.7.1 Mechanical splicing

In the mechanical splicing method, belt ends are joined by mechanical fasteners after they brought nearer together. There are many types of mechanical fasteners on the market today. Most work on the principle of clamping the two ends of the belt together between metal fasteners. Mechanical fasteners are manufactured principally in two styles, solid plate and hinged with a variety of attachment methods, including bolts, rivets and staples. They are fabricated from a variety of metals to resist corrosion and wear.

One end of the solid plate fasteners is installed on one end of the belt, and other end of the fasteners is installed on another end of the belt; thus, two ends of the belt are joined. In the case of hinged pin fasteners, a row of mechanical fasteners are physically attached to each belt end. Thus, attached fasteners to both ends are then meshed together and connected with a hinge pin. In both cases, it is essential to prepare straight and squared belt ends. Otherwise, tracking problems and belt edge damage can follow.

Solid plate fastener systems are ideal for larger pulley diameters requiring higher tension belt splicing. With no working movable parts, solid plate fastener systems will generally deliver long, trouble-free service life. They help to eliminate sifting of fine materials between plates.

Hinged fastener systems can be separated in order to remove, extend or clean belts, simply by removing the hinge pin. They can be spliced in the shop, and only the hinge pin has to be inserted on site. Hinged fasteners can be used to join belts of slightly different thicknesses.

When either style is appropriate, solid plate styles are preferred for longer life and to prevent sifting. Hinged fasteners are preferred on portable conveyors and on conveyors with smaller pulley diameters.

Conveyor belt fasteners can be divided into two major segments as light-duty and heavy-duty applications. Light-duty belts have a tension rating of 160 PIW (pounds per inch of width) or less, and heavy-duty belts have tension ratings over 160 PIW. Light-duty conveyor belt fasteners include stapled plate fasteners, common bar lacing, rivet fasteners, spiral lace and wire hook fasteners. Heavy-duty conveyor belt fasteners include bolted and riveted solid plate fasteners and stapled, bolted and riveted hinge fasteners. Figure 4.22 shows all the above types of fasteners.

When the belt is ready after splicing with mechanical fasteners, the belt is to be run lightly loaded first and then increase the load in stages until the belt is running

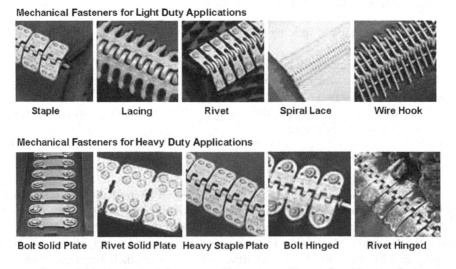

Figure 4.22 Mechanical fasteners. (Courtesy – Flexco Steel Lacing Co.)

satisfactorily with full load. The belt should be carefully monitored for at least one day's full working.

Splicing by using mechanical fasteners offers an economical, reliable and long-lasting belt splice method. Mechanical fasteners are a good choice for both new and older worn belts. Temporary splices, with mechanical fasteners, are very popular for conveyor belts, which are subject to frequent replacement, shortening or lengthening in operation, as in underground mining, or open cast mines, where shiftable conveyors are used. In some cases where very sticky materials such as damp coal fines are handled, the mechanical splice can actually be a benefit in belt cleaning by providing a periodic rapping of the belt cleaner to remove build-up. When the belt is spliced with fasteners, material may spill through the crevices of fasteners. To prevent this, one or two layers of tie-gum are applied over fasteners as a temporary measure which would be quite effective.

Mechanical splices can be installed relatively easily with only a modest amount of mechanical skills and tools by plant personnel, but as a consequence, they can easily be misapplied. Because of their comparative ease of installation, mechanical splices are often used in emergency repair situations, when a new piece must be added to an old belt or when a belt must be patched or a rip closed to restore the operation of conveyor.

Mechanical splicing with fasteners has the following advantages:

- Splicing can be done relatively faster
- Initial expenditure for making splicing is lower
- Take-up travel problems are minimized
- There is not much loss of belt
- Not much skill is required
- Because mechanical splice is visible, wear and deterioration is apparent and can be taken care of before a complete belt failure

The following are the disadvantages of mechanical splicing with fasteners:

- Compared with vulcanized splicing, mechanical splice with fasteners has lower joining strength
- Fine material spill from the fastener gaps
- Fasteners damage the pulleys and idlers
- Fasteners also damage lagging
- In hot service, fasteners retain heat and transmit it directly into the belt carcass
- As mechanical splicing produces rough surface, belt wipers sometimes catch on the fasteners and suffer damage

4.7.2 Vulcanization

Vulcanization is a complex process that yields a smooth splice with minimal risk of snagging, tearing and other harmful wear to the belt. Preparation of the straight and accurate squaring of the belt ends is crucial in a belt joining technique. Two types of vulcanization processes are hot vulcanization and cold vulcanization. Each process requires unique tools and an intimate knowledge of the rubber bonding process.

The hot vulcanization of multi-ply textile reinforced belts starts with stepping the two belt ends. Each side of the belt is stepped oppositely with the number of steps equal to one less than the number of plies of the belt. Thus, prepared belt ends are brought together as shown in Figure 4.23 and the fit is checked.

The later steps are as follows:

1. Cleaning the stepped belt thoroughly with benzene/toluene/trichloroethane so that traces of grease, oil, etc., are removed
2. Application of vulcanizing solution over the prepared splice area
3. Application of insulation gum strips over the splice area
4. Matching the belt ends
5. Filling in the cover recesses

In the case of solid-woven textile reinforced belts, triangular fingers are cut into the belt ends to be joined. These are brought together, leaving a gap between the fingers from each side. The gap is filled with paste, a reinforcing fabric placed above and below the fingers and then covered with a protective elastomeric material. Thus, formed splice is called finger splice. In steel cord reinforced belts, the rubber surrounding the steel cords is removed. The cords from each belt end are overlapped in a symmetrical pattern, and the intermediate gap is filled with unvulcanized rubber. Unvulcanized cover rubber is placed on either side of the cord line. Thus, formed splice is called overlap splice. Finger splice of solid-woven textile belt and three-step overlap splice of steel cord belt are shown in Figure 4.24.

Thus, prepared splice is cured in a vulcanizing press by the application of required temperature and pressure to form an endless loop. The joint made by the vulcanization must be protected from dust, and condensation during the making of the splice and elaborate screening may be essential to guarantee good result. A slight air pressure maintained inside the splicing enclosure helps to keep out the dust.

In cold vulcanization, the belt's layers are lapped with glue that provides a cure at room temperature. Cold vulcanization does not employ a vulcanizing press, but uses a bonding agent that causes a chemical reaction to splice the two belt ends together.

Whether a belt should be hot-vulcanized or cold-spliced is a matter of convenience. The development of cold splicing system is aimed at splicing conveyor belts by cold bonding without using heated vulcanizing equipment. The primary advantage of cold vulcanization is the saving in time and labour both in transport and assembly of the vulcanizing press and in curing time. Also where explosion due to electric sparks is to be avoided, as in some underground mines, cold splicing, without the use of explosion proof vulcanizing plates, is a cost-saving method. The cold splicing cement ensures

Figure 4.23 Four-ply textile stepped belt ends.

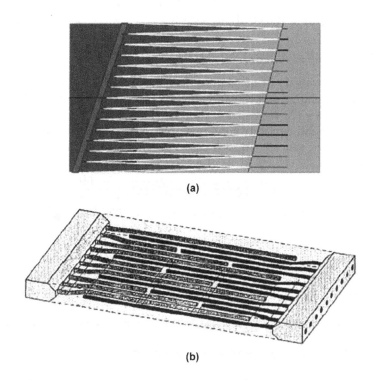

Figure 4.24 Finger and overlap splices of (a) solid-woven and (b) steel card belts.

excellent adhesion and high dynamic efficiency. Cold vulcanization needs more meticulous care and craftsmanship.

In case of any doubt whether for a particular application, hot vulcanization or cold splicing is better, it is better to make two test samples, one of each, and test for the strength of the joint, and then decide. Almost all new belts are manufactured and cured with the hot vulcanization process for better ply adhesion.

Cold splicing is not suitable for belts, on which hot materials are transported. For cold splicing, the ambient temperature should not exceed 80°C. High atmospheric humidity and severe dust conditions do have adverse action on splicing.

A vulcanized splice is the strongest and most reliable method of joining a belt and should be used wherever possible. It is less troublesome than fastening methods and results in longer belt life. The carcass of the belt is fully protected, and the surface is left smooth so that scrapers and brushes can be used more effectively and with less risk of damage to the belt than with a fastener joint.

The initial cost of a vulcanized splice is many times greater than that of a mechanically fastened splice. To insure sufficient take-up travel for accommodating both elastic and permanent variation in belt length, longer take-up travel must be provided in case of the vulcanized splicing method. On long centre belts and also on installations where the take-up is limited, it may be necessary to run in the belt with the fastener joint for a period of time. This ensures the removal of the initial elongation in the belt, before making the final vulcanized joint. Replacing or renewing a vulcanized splice can be time-consuming and costly, especially in emergency repair situations.

4.7.3 Mechanical vs vulcanized splicing

The speed and simplicity of mechanical splicing are advantageous over the vulcanized splicing. Depending on belt width and thickness, most mechanical splices can be finished in less than 1 hour. Vulcanization takes from 4 to 6 hours and also requires ideal temperature and humidity conditions. Worn belts and many older belts will not accept a vulcanized splicing. In such cases, splicing with mechanical fasteners is the only alternative.

Mechanical splicing requires simple instructions and portable application tools such as hammers, punches, wrenches or lacers. Hot vulcanization, on the other hand, requires large and heavy heating presses and special hand tools to split the belt plies and prepare the belt surface. Newer belt synthetics must also be carefully matched with vulcanizing solvents, cements and compounds.

In case of mechanically spliced belt, indication of possible failure can usually be detected during routine inspection. Replacement or repairs can be made at the following scheduled shutdown, thus preventing costly downtime. A deteriorating vulcanized splice, in comparison, is not easily detected. Catastrophic line failure can be the first indication of trouble once adhesion breakdown occurs inside a vulcanized splice.

Chapter 5

Idlers/Rollers

Rollers, which are in cylindrical shape, support the belt between two pulleys at both ends of the belt conveyor. As the belt movement imparts rotary motion to rollers, the belt moves linearly on supporting rollers without sliding. **Idler set** (or idler) generally means complete unit comprising roller(s) along with its mounting frame. Figure 5.1 shows single roller, mounting frame and three-roll idler set.

Conveyor idlers take the first place in terms of number present in a conveyor system. Cost-wise, it takes the second place next to the belt. There are many types of idlers, but regardless of the design, their main functions are as follows:

1. to provide trough shape for the belt to contain the material
2. to provide support for the belt with minimum resistance to the belt motion
3. to minimize the power needed to transport the material

An idler is a non-powered rotating mechanism which carries the belt. It rotates freely and neither receives nor transmits power to the belt. Hence, it is usually referred to as **idler**. But this concept is not fully valid today because this non-powered part is a powerful part as the total reliability and operational characteristics of conveyor depend on this. This also contributes to the maintenance cost, power requirement, belt performance and replacement, etc., resulting in boosting the overall productivity. Hence, they are termed as **rollers**, and the term **rollers** is used hereafter in this book.

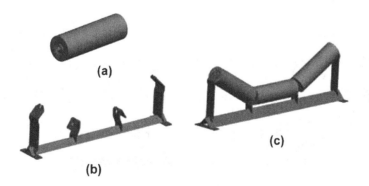

Figure 5.1 (a) Roller, (b) frame and (c) idler set.

5.1 Components of the roller

In general, a roller has the following basic components:

1. An outer shell or tube
2. A central axle or spindle or shaft
3. A pair of identical bearings (one on each side)
4. Suitable housings to accommodate the bearings
5. Sealing arrangement
6. Locking arrangements for the internal parts
7. Fixing arrangement of the roller to the frames

The components of a typical roller are shown in Figure 5.2.

5.1.1 Tube (shell)

Mainly, there are two types of steel tubes: (1) seamless tubes and (2) E.R.W. (electrical resistance welded) tubes for roller application.

Seamless tubes have a homogeneous wall without a weld or joint along its length. It is made by punching a hole through a red hot billet. Seamless tubes are basically strong and have uniform mechanical/chemical properties. These tubes are also free from age hardening. Hence, the seamless tubes should be normally necessary where corrosion, high internal pressure, etc., exist. Such applications do not warrant very close tolerances, in thickness and concentricity. But, because of the severity of the forging operations and other processes, the seamless tubes are much more expensive than the E.R.W. tubes.

E.R.W. tubes are manufactured by bending steel metal sheet and welding longitudinally. This welding is done by electrical resistance process, and hence, these are called E.R.W. tubes. These tubes could be identified by a thin line of welding seam that could be seen inside the tube. For mechanical applications and where the tubes are subjected to external loads as in the case of rollers, E.R.W. pipes are found to be best suited in view of the cost as well as the performance. While procuring E.R.W. tubes, it must be ensured that the concentricity of the tube and the wall thickness have been maintained to the closest tolerance, as any ovality in the tube would result in the jumpy movements of the conveyed material, heavy vibration of the rollers and subsequent damage to the bearings.

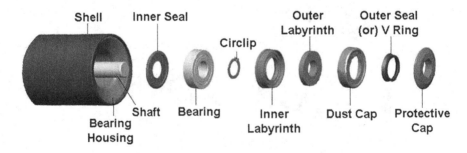

Figure 5.2 Components of a roller.

5.1.2 Bearing housing

Bearing housing is a metallic housing, obtained by stamping, forging or machining depending on the working application of the roller. The bearings and other internal parts are fitted in these metallic housings, which are either welded to the roller shell or press-fitted. The bearing housing should fix snugly with the shell to take care of the static and dynamic operating conditions. Similarly, the housing should have a close tolerance for accommodating the outer race of the bearing, without any slip. The bearing housings should be welded to the shell, simultaneously at both the ends to prevent any **eccentric** running of the roller. These housings are made of cast iron in the majority of the cases. But, due to many deficiencies, they are progressively being replaced by sheet metal pressed housings. Some of the deficiencies are also being overcome by improved-quality cast alloy housings, for special applications.

5.1.3 Shaft

The shaft is a long solid cylindrical section, is precision-machined to house the bearings and has bevelled ends to facilitate mounting of the bearings. The dimensions of the shaft are very critical, and any variation would affect the assembly of the internal parts. Hence, during manufacture, care has to be taken for maintaining the required tolerances. The design of shaft is critical to the life of the bearing. It is possible to get maximum life out of bearings by stepping down the shaft and reduce the deflection and use lesser size bearings. Bright steel bars are preferred for roller shafts as these could be drawn easily to the final dimensions and final machining is also considerably reduced.

5.1.4 Bearings

The life of a roller is mainly governed by the life and performance of the bearings, although other factors such as effective sealing, quality of lubricant, spacing of rollers, load rating and external friction factor do influence. In the early stages, using bush bearings for rollers was very common, but now mostly anti-friction bearings are used.

A bearing can be chosen specifically and individually for each and every application from the wide variety available as shown below:

Ball Bearing: deep groove; angular contact; self-aligning
Roller Bearing: spherical roller; cylindrical roller; tapered roller
Thrust Bearing: single-acting; double-acting, etc.

Ball bearings are better suited for conveyor roller application.

5.1.5 Sealing

The wear of the bearings is influenced to a large extent by the quality, type and life of the sealings. The sealings should be specially designed to keep dust, dirt and water away from the bearings and at the same time prevent the lubricant from flowing out. The life of the bearing is highly dependent on the efficacy of the sealings.

Basically, there are two different types of sealings

1. Non-contact or labyrinths and V-rings
2. Contact or oil seals

Non-contact seals or labyrinths have the advantage that they are not subjected to direct wear as the contact seals. They are simple in construction, are easy to manufacture in bulk, give considerably longer life and are economical. The friction factor in the case of non-contact seals is much lower than that of contact seals.

Labyrinth can be described as the heart of the sealing arrangement in a roller. The word labyrinth itself indicates that it presents a circuitous and torturous path for the dust and other contamination when it is filled with grease. The basic principle of a labyrinth seal is based on the geometric shape of the labyrinth seals which causes some turns of the contaminant on its way to penetrate the seal. The major factor for the efficiency of a non-contact seal is the centrifugal force caused by rotation and applied to the contaminant to throw it radially away before penetrating the seal. At higher peripheral speed, an air barrier is built inside the seal keeping different types of contamination (e.g. dust or liquids) out.

There are single-, double- and multiple-grooved labyrinths depending upon the duty conditions. The depth, width and tolerances provided for the grooves are critical and require precise machining. Any mismatch of the inner and outer rings would increase friction and heat and ultimately result in the failure of the quality of the lubricants.

V-ring seals are axial shaft seals and utilize centrifugal slinger action to provide effective protection against dry and wet contamination. The V-ring is an all-rubber seal and is mounted directly on the shaft by hand and seals against a counterface. The V-ring, designed with a long, flexible lip, can act as a face seal, a lip seal or slinger. The V-ring seal has three parts: the body that clamps itself on the shaft, the conical self-adjusting lip and the hinge. The elastic body holds itself in place on the rotating shaft while dynamic sealing takes place where the lip is in axial contact with the counterface. The counterface should be metal and can be the end of a gearbox housing, a washer and a suitable steel stamping. Because of the peripheral speed, dirt and liquids are removed and will not attach to the sealing lip.

Contact or **lip type seals** are also used for conveyor rollers. A good contact seal is more expensive than a non-contact seal. Contact seals for particular applications may be more effective than non-contact seals. The efficacy of contact seal would depend on many factors such as quality of the clad material, quality of the rubber lip and springs.

5.1.6 Circlips and locking arrangements

Circlips are a type of retaining rings or fasteners that take the form of a flexible, open-ended ring, made from metal and used for holding structural components on the shaft. Fundamentally, there are only two types: internal, which are typically used to hold a seal or bearing in its bore by fitting inside a cylindrical bore or housing and push outwards, and external, which are used to locate components on shafts by fitting around a shaft and pressing on it. These are not load-carrying members. These are selected based upon experience to suit the duty conditions. Such retaining or fastening components are, cost-wise, negligible, but performance-wise, these are critical parts. Poor-quality and inadequately designed circlips may result in unexpected failure of the rollers.

5.1.7 Fixing arrangement of the roller to the frame

Fixing arrangements for different rollers are described in Sections 5.2.3 and 5.2.4.

5.2 Types of rollers

There are two basic types of belt conveyor rollers:

1. **Carrying rollers**, which support the loaded run of the belt
2. **Return rollers**, which support the empty return run of the belt

Within these two types of rollers, there are a number of different designs of roller sets that have developed as a result of particular application, the need to minimize the stress imposed onto the belt and the environment in which the conveyor operates.

5.2.1 Carrying rollers

Function-wise, the carrying rollers are generally classified as follows:

1. Flat belt rollers
2. Troughing rollers
3. Impact rollers
4. Transition rollers

Flat Belt Rollers: Flat belt rollers consist of one horizontal roller to support the belt. Material spillage, capacity and load centring can be a problem with this type of roller. Flat rollers are often used on feeders that are known to be difficult to train and seal (Figure 5.3).

Troughing rollers: Troughing roller set consists of three, five or seven rollers but usually consists of three rollers. Central roller is arranged horizontally, while the two end rollers are fixed at the required angle to form the necessary troughed cross section. Theoretically, trough of a belt can be formed by combining rollers of any length, but for particular application, material, speed, etc., the lengths of central and end rollers should be calculated and selected for forming the most advantageous trough in terms of volume and capacity.

Figure 5.3 Flat belt roller.

For carrying bulk materials, the troughing angle could be 15°, 20°, 25°, 30°, 35°, 40° or 45° as the same is limited to a major extent by the type of belt. For example, if cotton–cotton belt is used, depending on the number of plies, the troughing angle could be between 0° and 25°. In the case of cotton–nylon belts, it could be slightly higher. The maximum troughing angle could be as much as 60° at feeding points, in the case of steel cord belts. Generally, as there is no weft member in steel cord belt, the belt is quite flexible.

The troughing angle ensures that the carrying belt maintains the same cross-sectional area throughout the carrying strand, so that the load-bearing capacity of a particular conveyor belt is the same along the full length of the conveyor. The material loaded to the maximum capacity at the feeding point will not fall off throughout the rest of the belt.

As the troughing angle increases, the cross section and the carrying capacity (tonnage) also increase up to a certain degree but start decreasing afterwards. An increase in troughing angle in turn increases the strain on the belt and rollers. The carrying capacity of a system can often be altered by changing the length of the centre roll to widen or narrow the trough.

Although 20° troughing rollers were once standard, 35° and 45° troughing rollers, called deep-trough rollers, are now more common. The benefits of these deep trough rollers are as follows:

*   Greater cross section for more tons per hour
*   Less cover wear per ton of material carried
*   Use of narrower belt is possible – for savings on belt and hardware
*   Reduced potential for material spillage
*   With most materials, deep trough allows for more distance between rollers

Three-roll troughing roller sets are made in two general styles as **in-line** and **offset**. **In-line** troughing roller set consists of three equal-length rollers mounted on the roller frames with the centre lines of the three rollers aligned (Figure 5.4). For a given width of the belt, end-roll inclination and material surcharge angle, three equal-length troughing rollers form the belt in the best troughed shape to carry a maximum load cross section.

In **offset** troughing roller set, the centre lines of the inclined rollers are in one plane, while the centre line of the centre roller that is horizontal is located in a plane alongside the plane of the inclined rollers (Figure 5.5). Offset roller set should be installed in such a way that the moving belt contacts the centre roller before it touches the wing rollers. Offset roller set prevents roller junction-type belt failure by offering improved

Figure 5.4 In-line troughing roller set.

Figure 5.5 Offset troughing roller set.

belt support. It is easy to replace centre roller of offset roller set. The most common use is in installations with minimal clearances. Offset roller sets are popular in the grain industry, where very thin belts are used, and in underground mining, where low head room is a problem. Offset rollers offer more protection against roller junction fatigue than the in-line type.

A **picking roller** is a specialized in-line troughing roller furnished with a long horizontal roller and two shorter inclined rollers (Figure 5.6). The short inclined rollers provide a slight trough to minimize spillage, while the longer horizontal roller allows the load to be spread. This type of trough provides a large flat area to carry material and allows easy inspection or picking and sorting of the carried material.

Impact Rollers: Impact rollers, sometimes referred to as **cushion rollers**, are specially designed to be used in the feed zone to protect the belt from impact damage. Wherever material is fed onto a conveyor belt, impact rollers are installed beneath the troughed belt over the full feeding length. They should be located and spaced in such a way that the principal portion of the lumpy material flow engages the belt between the supporting rollers rather than over any one of them. The cushioning ability of impact rollers allows the energy associated with the impact to be absorbed more efficiently and with much less detrimental effect to the belt. To support flat belts, rubber-cushion impact rollers are made (Figure 5.7).

One type frequently used for troughed belt consists of a three-roller assembly, each roller being made of spaced discs of rubber or other resilient material (Figure 5.8). These discs or rings flex under impact to absorb the energy of the materials fed onto the belt. The rubber rings can be solid rubber rings or tyre-type hollow rings. Tyre rings have an air ventilation hole to prevent damage due to excessive impact. These hollow tyre-type rubber rings absorb shock better but are more expensive. Some manufacturers use impact pads (also called impact cradles) at the feeding sections instead of rubber rings (Figure 5.9).

Figure 5.6 Picking roller.

Figure 5.7 Flat impact roller.

Figure 5.8 3 Roll impact roller.

Figure 5.9 Impact pads.

Transition Rollers: Transition rollers are provided on the conveyors where the belt changes from flat to trough shape while passing out from the pulleys and vice versa. These are similar to normal troughing rollers, but the troughing angles are gradually increased/reduced in two to three stages, to effect a smooth transition of the belt. As the belt leaves the tail pulley and moves into the loading zone, the belt edges are elevated forming the trough shape. If the belt directly takes full trough form from the tail pulley, the belt edges would be over-stressed and damage occurs.

Transition distance is the distance from the centre line of the first fully troughed roller to the centre line of either the head or tail pulley. Transition distance varies with the amount of troughing required, belt thickness, construction of the belt and the rated tension of the belt. If the transition distance is too short, the edge of the belt can be over-stretched. This will adversely affect the load support and belt life. A general rule of thumb for minimum transition distances is to provide at least the following:

1.0 × belt width	for 20° troughing rollers
1.5 × belt width	for 35° troughing rollers
2.0 × belt width	for 45° troughing rollers

Depending on the transition distance, one, two or more transitional troughing rollers can be used to support the belt between the terminal pulley and the closest fully troughed roller. The number of transition rollers depends on the trough angle of the conveyor. In the case of a 45° trough angle, two- or three-transition-roller sets would be used at either end of the conveyor. Figure 5.10 shows the transition distance and arrangement of transition rollers.

It is good practice to install several transition rollers with intermediate angles supporting belt to work up gradually from a flat profile to fully troughed contour at the tail end of the belt conveyor. In Figure 5.10, two transition rollers with 20° and 35° angles are arranged. At the head end of the belt conveyor, the belt changes from trough shape to flat shape by the time the belt comes to the head pulley. Here again, two transition rollers with 35° and 20° angles are arranged.

As the stability of the belt is important, transition roller closest to the terminal pulley is installed in such a way that the top of the pulley and the top of the centre roll are in the same horizontal plane.

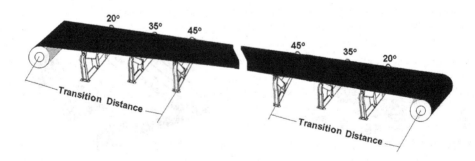

Figure 5.10 Transition distance and arrangement of transition rollers.

5.2.2 Return rollers

The mass of the return belt is the only load that return rollers are required to support. As such, return roller sets are spaced at two to three times the spacing of their equivalent carrying-side roller sets. Return roller sets usually have one or two rolls per roller set.

The following are different types of return rollers:

1. Flat belt return rollers
2. Inverted troughing rollers
3. Spiral rollers (self-cleaning return rollers)
4. Two-roll V return rollers

Flat Belt Return Rollers: Return rollers provide support to the belt after it has unloaded its material and is on its way back to the tail pulley. These rollers normally consist of a solid, single horizontal roller mounted to the underside of the conveyor stringer (Figure 5.11). As the dirty side of the belt comes in contact with the return rollers, these rollers are also provided with rubber discs or rings to help better cleaning of the belt and to prevent coating of the conveyed material on the rollers (Figure 5.12).

Inverted Troughing Rollers: To take care of the wandering of the return belt and to guide the belt, inverted troughing rollers are provided near the tail ends. This would prevent, to some extent, the lining-out of the return belt and reduce the edge damages in the belts.

Spiral Rollers (Self-Cleaning Return Rollers): If the carrying material adheres to carrying surface of the belt and if such material is abrasive, it may wear the shell of the return rollers or it may adhere to the return rollers, which causes misalignment of the return run of the belt. In such cases, spiral rollers are provided on the return side of the belt specially for cleaning the return belt. When the return belt passes over these rollers, spirals provided on the rollers effectively clean the belt by their outward spiral movement. Figure 5.13 shows spiral roller in use. Spiral roller sometimes is erroneously called a **belt cleaning roller**. Even though such rollers do track off material adhering to the belt surface on the return run, they do not constitute the belt cleaning devices.

As the metallic spirals are always in contact with the belt, care should be taken to see that the damaged/worn-out spiral rollers are replaced immediately, as otherwise they are likely to cause extensive damage to the belt. These spiral rollers can be used for belt travel in one direction only.

Disc roller, shown in Figure 5.14, consists of steel or rubber discs instead of spiral, is used for the same purpose as that of spiral roller and can be used for belt travel in either direction.

Two-Roll V Return Rollers: With the increased use of heavy, high-tension fabric and steel cable belts, the need for better support and training has resulted in the development of V return rollers. A basic V return roller consists of two rolls, each tilted

Figure 5.11 Flat belt return roller.

Figure 5.12 Return roller with rubber discs.

Figure 5.13 Spiral roller.

Figure 5.14 Disc roller.

Figure 5.15 Fixed two-roll V return roller.

Figure 5.16 Garland two-rubber-disc-roll V return roller.

approximately 10°–15° angle. V return rollers have some training effect on the belt, while allowing greater roller spacing because of its increased load rating. These can be either steel-roll or rubber-disc designs. Field experience has shown that the steel roll is preferable because discs tend to wear out rapidly.

Two-roll V return rollers are either of the fixed two-roll V return roller as shown in Figure 5.15 or of garland (also called as catenary or suspended) two-rubber-disc-roll V return roller as shown in Figure 5.16.

Depending upon the method of fixing of the rollers on the conveyor structure/brackets, the rollers can be classified into two categories:

1. Fixed type
2. Garland type (catenary or suspended)

5.2.3 Fixed-type rollers

In this type, the rollers are fixed on the brackets as shown in Figures 5.4 and 5.5, which are in turn installed on the stringers of conveyor structure. There are mainly two types of conveyor structures, viz. floor-mounted structure and roof-hung structure. Figure 5.17 shows floor-mounted structure with in-line 3-roll carrying roller set

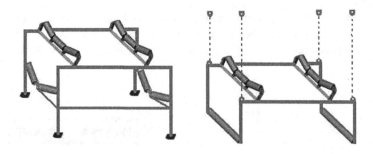

Figure 5.17 Floor-mounted and roof-hung conveyor structures.

with 2-roll V return roller set and roof-hung structure with in-line 3-roll carrying roller set with single return roller. The vertical members called stand or post are provided at regular interval to support the stringers.

Fixed-type rollers are classified into two types as rollers with outside bearings and rollers with inside bearings.

Rollers with Outside Bearings: In this type, the bearings are mounted on separate pedestals and the shaft rotates integral with the shell. These are also known as **live shaft rollers**. Though these types of rollers were used in the earlier days, they are not used in large installations nowadays due to the advantages of dead shaft roller.

Rollers with Inside Bearings: Here, the bearings are inside the shell and the shell rotates while the shaft is stationary. Hence, these are called **dead shaft rollers**. The roller shell is fitted with bearing housings at both ends. In this design, the advantages are as follows:

1. Since the inner race of the bearing is fixed onto the shaft, there is point loading, and hence, assembly defects and minute variation in the alignment do not affect the performance of the rollers
2. The bearings are always perpendicular to the shaft irrespective of the misalignment of the shaft; i.e., it is completely independent of its support
3. For arduous conditions, a roller of this type is preferred, as, in this design, all the three rollers of a troughing set can be mounted in one line without permitting too large a gap between the end and central rollers. This reduces the **pinching** of the belt, which is one of the causes for ply separation in a textile belting and objected to, by belting manufacturers.
4. These types of rollers can be easily formed into garland rollers
5. For calculating the power required, the rotating weight alone is to be taken, which will result in less power

The manufacturing methods of these types of rollers are also comparatively much simpler, economical and generally preferred for mass production techniques.

5.2.4 Garland (catenary or suspended) rollers

In this design, the rollers are not installed on supports, but are connected together by means of linkages and formed into a garland. Such a garland can be made of either 2, 3, 5 or 7 individual dead shaft rollers.

Two- and three-roll garland rollers are used as carrying rollers, whereas for return rollers, only two-roll garlands are used. Similarly, 5- and 7-roll garland rollers are used at the impact sections and tail ends in view of their flexibility and higher dampening effect of impact stresses.

Figure 5.18 shows three-roll garland rollers with three different garland hanger hooks, chain links and rod ends. Figure 5.19 shows different types of garland rollers.

The garland roller set is hung from the two pivots provided on the side stringers of the conveyor structure so as to form the necessary troughed shape. Figure 5.20 shows the conveyor structure with 3-roll carrying garland roller sets and 2-roll return garland roller sets.

Figure 5.18 Three-roll garland rollers with three different types of hangers.

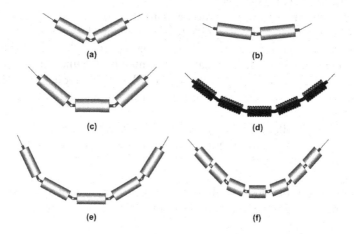

Figure 5.19 Different garland rollers. (a) Two-roll troughing rollers, (b) 2-roll return rollers, (c) 3-roll troughing rollers, (d) 5-roll impact rollers, (e) 5-roll troughing rollers and (f) 7-roll troughing rollers.

Figure 5.20 Conveyor structure with garland rollers. (Courtesy: MELCO Conveyor Equipment (Pvt) Ltd.)

In comparison with the fixed type of rollers, the garland rollers have the following specific advantages:

1. Better guidance of the belt due to the transverse flexibility of the garlands.
2. Self-centralizing of the material on the belt and, hence, spillage are avoided. The garland rollers are also flexible to the **out-of-centre** loading of the materials. In the case of a three-piece garland, the garland itself undergoes a change in the troughed shape when the load is unsymmetrical. Under such conditions, the central roller can move from the normal position till it is in line with one of the side rollers.

   In the case of a five-piece garland, which is preferred for impact sections of broader belt, when the load is on the centre, the central roller itself can move downwards until the side rollers are stretched. Each garland roller can, when struck by a lump, move a little out of the way, i.e. forward, backward, downward or sideways. This causes the garland to be deformed in their trough and thus assists in dampening the impact energy and helps in reducing partly the kinetic stress of the impact load.
3. Comparison between rigid roller stations and garland system showed that the impact in rigid rollers was two to four times more than the garland system. The impact could be further reduced with rubber rings.
4. Adaptability of the troughed shape according to the shape of material.
5. Garland rollers offer less of a jolt to traversing loads than standard rollers.
6. When transporting lump materials, there is a considerable reduction in the dynamic stress on the bearing at loading points.
7. The stress of the belt is reduced at the loading point, by nearly 40–50% of that of rigid roller sets, thereby increasing the life of the belting.
8. Increased life of the rollers due to lesser vibration.
9. Consequential reduction in the maintenance cost.
10. Quieter running of the rollers.
11. Reduction in structural weight by approximately 10%–15%.
12. Reduction in the assembly cost.
13. In the case of a failure of one roller in a garland set, in certain special designs, the particular roller set could be released, even while the belt is in operation. This is accomplished by a special device called the quick release lever. The replacement of the failed roller could be taken up during the next planned stoppage, whereas in the case of fixed type of rollers, such a failure may lead to the jamming of the roller, and in turn damage to the belting. There is also the possibility of fire hazards.

Garland rollers at the impact section have been generally found to be very advantageous. Garland impact rollers can be provided with quality rubber rings to dampen the impact force. But, rubber rings always require constant attention and replacement. This would increase the cost of maintenance. At the same time, the shells of the impact rollers if not provided with cushion rings would get deformed under heavy impacts and also by collision with each other during operation especially with high-speed conveyors.

5.2.5 Rollers for training or tracking the belt

Well-designed, carefully constructed and maintained belt conveyors will continue to run with proper alignment without the need for special rollers such as training or tracking rollers. However, there are transient conditions that may cause belts to become misaligned despite all efforts to assure proper installation and maintenance. Many times, it is observed that the belts of the conveyors, instead of running true and straight, run to one side. A few of the reasons for such wandering or snaking of the belt to the left or to the right are as follows:

• Misalignment of conveyor structure and components in the centre line
• Off-centre feeding of the material
• Accumulation of fugitive material on rollers and pulleys
• Improper belt splicing (not squaring of belt ends prior to splicing)
• Defective manufacture of the belt (bowed belt)
• Non-uniform wear of lagging of the pulleys
• Structural damage caused by inattentive heavy equipment operators
• Ground subsidence
• Conveyor is subjected to lateral winds
• Conveyor is subjected to rain, ice, snow or heat from one side

To correct this, training or tracking rollers are provided in sufficient numbers in a belt conveyor at regular intervals. Training or tracking rollers are self-aligned rollers that are able to detect belt misalignment and automatically realign the belt. Hence, they are also termed as self-aligning rollers and/or self-aligning training rollers.

Self-Aligning Training Rollers: Roller set used for training the belt normally has three rolls. This 3-roll self-aligning training roller set (Figure 5.21) is a special roller set mounted on swivelling frame which is centrally pivoted on stationary frame. Usually, a vertical guide roller, which should be normal to the belt edge, is provided at each end of the roller frame. If the belt sways to one side, it presses against a lateral vertical guide roller. Due to the braking of the end carrying roller, the self-aligning training roller frame is turned forward about the pivot point in the direction of the belt run. The whole roller set takes an oblique position forcing the belt back into its normal position. Then, the pressure on the vertical guide roller disengages, and the braking on

Figure 5.21 Self-aligning training roller.

the end carrying rollers, is released. The self-aligning training roller set now retracts to normal position. Thus, the self-aligning training rollers train the belt and protect the belt edges from damage caused by misalignment.

Self-aligning training rollers are of positive action type for belts operating in one direction and actuating shoe type for two directional operations (reversing belts). The effectiveness of these self-aligning training rollers depends on the belt tension. In general, the greater the belt tensions, the less effective the self-aligning training rollers are.

Usually, self-aligning training roller sets are mounted at intervals of approximately 50 m in long conveyors. Improper design and positioning of the vertical guide rollers could cause heavy damage to the belt due to wedging of the belt below the roller and preventing free movement of the belt. These guide rollers contact the belt edge when the belt is mistracking and can cause edge damage in the case of severe or continued mistracking.

Self-aligning training rollers are especially useful near the head and tail pulleys, since these are the loading and discharge areas where the belt should always run on centre. One or two self-aligning training rollers at intervals ahead of each pulley will guide the belt into the pulley properly. When installing, the centre roll should be elevated 12–18 mm higher than the rolls of the adjacent rollers. This helps to ensure that consistent belt pressure is maintained on the rollers to allow it to function properly. This is applicable to both carrying- and return-side self-aligning training rollers.

Return Training Rollers: Return training rollers correct the belt mistracking as the belt returns to the tail pulley. Return training rollers are favoured as more hangers and enclosures at the return side can endanger a mistracking belt. As the return side is the low-tension side of the conveyor, the belt is more receptive to training on this side.

There are two basic designs for this type of roller. The first one is very similar to the carrying training roller, as it has a centre pivot bearing on the roller frame and two belt edge rollers (Figure 5.22). This type of return training roller uses a double-row spherical bearing mounted inside the centre of the roller. As soon as the belt starts to run off centre, this training roller acts immediately to correct and recentre the belt.

The second design consists of a multiple-axis bearing located inside the roller (Figure 5.23). As the belt begins to mistrack, the force of the movement causes the bearing to pivot and tilt, so the roller is positioned to correct the mistracking. A chevron ribbed cover provided on some training rollers of this type assists in grabbing and steering the belt back onto the track.

V return rollers (Figure 5.15) are becoming more popular on long, high-tension conveyors. These rollers create better belt-to-roller contact than flat rollers, so training becomes more controllable. However, they are generally more expensive and require somewhat more maintenance than conventional return rollers.

Figure 5.22 Return training roller.

Figure 5.23 Multiple-axis training roller. (Courtesy: Cleveland Oak.)

It is to be noted that conveyor operations should not depend on these training rollers to overcome gross misalignments of conveyor structure. Training rollers are not intended to continually alter the path of belt movement. Continuous working of this roller indicates a more serious problem with the belt tracking which should be identified and corrected.

5.3 Roller spacing

Roller spacing has a dramatic effect on both the support and shape functions of the roller. Obviously, roller spaced too far apart will not properly support the belt or enable it to maintain the desired profile. Spacing rollers very close together will provide sufficient support and profile but may be unwise economically. Normally, the supporting rollers are spaced near together, 3.5–5.0 ft apart according to the amount of material conveyed, to limit the sagging of the belt.

Any conveyor belt is expected to stretch or elongate, which causes sagging of the belt. Belt sag is the vertical deflection of a conveyor belt from a straight line between rollers. Belt sag is usually expressed as a percentage of the centre spacing of the rollers. If q is the normal distance from top of the roller to the lowest level of the belt between the two successive carrying rollers and s is the distance between two successive carrying rollers, then the belt sag between the rollers is defined as q/s as shown in Figure 5.24.

Normally, belt sag (q/s) ranges 1%–2%. Belt sag can be kept to a maximum of 2% with roller spacing at four-foot intervals along the carrying side. By increasing the tension, the roller spacing can be increased and still maintain the 2% maximum sag. In the load zone, the belt sag should be limited to 1%. If there is too much sag, the load will shift as the belt carries it up over each roller and down into the valley between. This shifting will increase belt wear, spill materials and require excessive conveyor drive power.

Belt tension also regulates roller spacing. A highly tensioned belt requires fewer belt support structures than a belt with a lower tension. Return roller spacing is determined by the weight of the belt since no other load is supported by these rollers. Return rollers may be twice as far apart as the carrying rollers and may be simple rollers, whatever the design of the carrying rollers.

Some conveyor systems have been designed successfully utilizing extended roller spacing and/or graduated roller spacing. Extended roller spacing is simply greater than normal spacing. This is sometimes applied where belt tension, sag, belt strength and roller rating permit. Advantages may be lower roller cost (fewer used) and better belt training.

In the case of graduated roller spacing, the rollers are spaced to reflect belt tension and sag requirements. Where tension is lowest, normally at the tail pulley and the loading zone, the rollers should be spaced close together. Where tension is high, as it

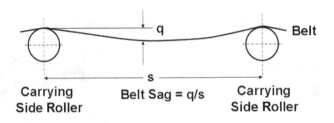

Figure 5.24 Belt sag.

usually is near the head pulley, the rollers should be spaced far apart. This graduated spacing method can both save power and reduce belt wear at the same time. In addition, fewer rollers are required for a given conveyor.

Areas that require additional support will need closer roller spacing. For example, at loading points, the carrying rollers should be spaced to keep the belt steady and to hold the belt in contact with the rubber edging of the loading skirts along its entire length. Careful attention to the spacing of the carrying rollers at the loading points will minimize material leakage under the skirtboards and, at the same time, will also minimize wear on the belt cover.

When the belt changes from the troughed shape to a flat profile over a pulley or vice versa, the belt edges are stretched and tension is increased at the outer edges. If the belt edge stress exceeds the elastic limit of the carcass, the belt edge will be stretched permanently and will cause belt training difficulties. On the other hand, if the troughing rollers are spaced too far from the terminal pulleys, spillage of the load is likely.

5.4 Belt alignment

A belt conveyor must be designed, constructed and maintained so that the belt consistently runs centrally on its mechanical system of rollers and pulleys. To accomplish this, the following conditions must prevail:

1. All rollers must be in line, square and level transversely
2. All pulleys must be in line, with the pulley shafts parallel to one another and at 90° to the centre line of the belt
3. The material must be loaded centrally on the belt
4. The belt must be straight and properly spliced
5. The supporting structure must be straight and level transversely

If, after these conditions have been met, the conveyor belt persistently runs to one side, the following two corrective measures with regard to rollers can make the belt run centrally:

1. Some of the rollers can be skewed so that the horizontal roll is at a slight angle to the centre line of the belt
2. Training rollers can be installed to replace either troughing or return rollers

The aligning action of the belt training rollers depends upon free movement of the conveyor belt and of the training rollers. Hence, proper cleanliness and maintenance are essential for satisfactory results.

5.5 Roller maintenance

Dirty or under-lubricated rollers can cause serious problems with the belt, as well as with the power costs. Material build-up can move belts offline to damage edges, and bearings can jam or freeze from material build-up. Jammed rollers drain system power and, if the shells are scored badly enough, can damage the belt cover.

Thus, rollers are subjected to various factors and conditions of working. Hence, inspection of rollers and lubrication of roller bearings are essential. Over-lubrication causes many more roller failures than under-lubrication. Additionally, the grease can accumulate inside the roller or on the belt and the structure. Surplus grease inside the roller unbalances it and increases power requirements, while grease on the belt can attack the rubber cover of the belt. As a result, while bearings requiring lubrication are still popular, the rollers employing a sealed-for-life bearing requiring no lubrication are designed and are now preferred.

Constant trials are being made to improve the internal parts of the roller. One of the recent developments is the design of seize-resistant bearings. Another recent development is good quality of grease with special additives so that it could stand without deterioration for nearly one lakh hours. Yet, another development is bringing down the internal parts from 10 to 12 numbers to the barest minimum, say 4 to 5 numbers. The design of sealing arrangement is undergoing phenomenal changes. To reduce the weight of the components, efforts are on to use more and more synthetic materials such as nylon and polyurethane.

Figure 5.25 shows belt conveyor with all types of rollers and pulleys, feed and discharge chutes and drive motor installed on the structure.

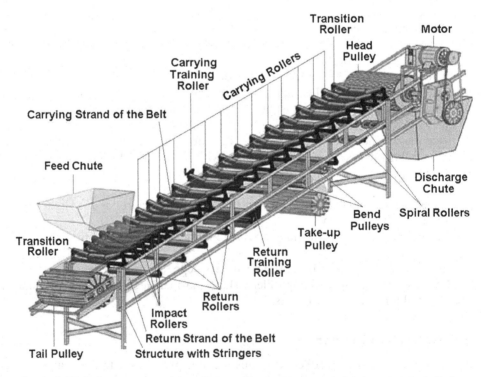

Figure 5.25 Belt conveyor with rollers, pulleys, chutes and drive motor installed on structure. (Courtesy: Rexnord Corporation.)

Chapter 6

Pulleys, shafts and drives

6.1 Pulleys

Pulleys, in a belt conveyor, change the direction of belt in a vertical plane and form an endless loop for continuous operation. Pulley (also known as drum) and shaft together form a composite structure whose operating characteristics are mutually related. Conveyor pulley construction has progressed from fabricated wood, through cast iron, to present steel fabrication. Pulleys are manufactured in a wide range of sizes. The most commonly used conveyor pulley is the standard steel pulley. Figure 6.1 shows lagged steel pulley with their components.

Drum or Shell: The drum or shell is fabricated either from a rolled sheet of steel or from hollow steel tubing. This shell of the pulley is in direct contact with the belt. Hence, the diameter and face width of the shell depend on the rating and width of the belt used.

Diaphragm Plates: The diaphragm or end plates of a pulley are circular discs welded to both the sides of the shell to strengthen the shell. These are fabricated from thick steel plate and are bored in their centre to accommodate the pulley shaft and the hubs for the pulley locking elements.

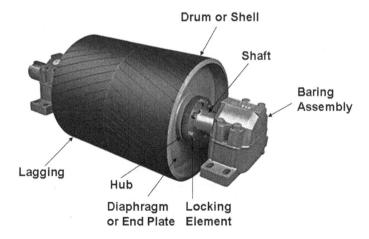

Figure 6.1 Components of the pulley. (Courtesy: ABB Motors and Mechanical Inc.)

Shaft: The shaft is a central cylindrical steel rod of the pulley designed to accommodate all the applied forces from the belt and the drive unit with minimum deflection. It is locked to the hubs of the end discs by means of locking elements. In the case of a shaft of drive pulley, the shafts are often stepped to accommodate different diameters of bearings and the drive attachment to the same shaft.

Locking Elements: Locking elements, which are high-precision items, when fitted over the shaft and into the pulley hubs, attach the pulley firmly to the shaft concentrically by tightening the screws around the locking element.

Hubs: The hubs, fabricated and machined housings, are welded into the end plates.

Lagging: Lagging is covering the pulley surface with rubberized material by vulcanization. Lagging is necessary to improve the friction between the conveyor belt and the pulley in order to improve the torque that can be transmitted through a drive pulley. Improved traction over a pulley assists training of the belt. In the figure, lagging with herringbone grooves is shown.

Bearing: Bearings used for conveyor pulleys are generally spherical roller bearings and self-aligning relative to their raceways. They are chosen for their radial and axial load supporting characteristics. The bearings are housed in plummer blocks that are bolted to sole plates. Sole plates are welded to the structure and incorporate jacking screws to enable the pulley to be correctly and relatively easily aligned.

Conveyor pulleys are designed for use on belt conveyor systems as a means to drive, redirect, provide tension to or help track the conveyor belt. Pulleys in a belt conveyor are primarily of two types, i.e. drive pulleys and non-drive pulleys. Drive pulley transmits the power to the belt in addition to supporting the belt, changing the direction of the belt in vertical plane and forming an endless loop for continuous operation. Non-drive pulley has the same function as that of drive pulley except that it will not transmit the power. Drive pulley has the live shaft, whereas non-drive pulley can have live or dead shaft. Pulleys are broadly classified into four types:

1. Flat- or straight-face pulleys
2. Spiral drum pulleys
3. Crowned pulleys
4. Wing pulleys

6.1.1 Flat- or straight-face pulleys

Pulleys of this type have flat face (Figure 6.2) and have less or no control on slippage of the belt. Straight-face pulleys are recommended for all installations using reduced-ply, high-modulus, low-stretch belts, such as those with a carcass of either steel cables or high-strength textile plies. These pulleys achieve the most conveyor belt contact, and because of this, they are the preferred configuration choice of pulleys in drive positions.

Figure 6.2 Flat-face pulley. (Courtesy: ABB Motors and Mechanical Inc.)

Figure 6.3 Spiral drum pulley. (Courtesy: ABB Motors and Mechanical Inc.)

6.1.2 Spiral drum pulleys

Spiral drum pulley (Figure 6.3) is formed by a pair of steel flat bars helically wound around a drum, having both begun from the opposite extremities of the pulley and meeting in the middle. There is a uniform pitch to these helices, and thus, material is uniformly discharged. Rotation of the pulley automatically starts the cleaning action discharging the material to the sides of the conveyor. Unique design of the spiral pulley reduces the possibility of material build-up between the belt and pulley where a self-cleaning wing pulley is not practical. It also reduces belt wear and misalignment.

6.1.3 Crowned pulleys

Pulleys of this type have crowned face (i.e. large diameter at the centre than at the edges), which is effective in centring a belt on the pulley if the approach to the pulley is an unsupported span that is unaffected by the steering action of rollers.

The three types of pulley crowns available are curve crown, taper crown and trapezoidal crown (Figure 6.4).

Curve Crown: Curve-crowned pulleys have a long, flat surface in the centre of the pulley with the ends curved to a smaller diameter. They provide enhanced belt tracking as compared to trapezoidal crown, an even wear surface and lengthened belt life.

Taper Crown: Face of the conveyor pulley is tapered with the high point located at the centre and tapering towards each end. They provide belt tracking capability, but the peak at the centre causes wear and stretching at the centre of the belt, decreasing belt life.

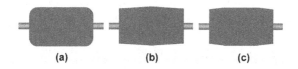

(a)   (b)   (c)

Figure 6.4 Crowned pulleys. (a) Curve crown, (b) taper crown and (c) trapezoidal crown.

Trapezoidal Crown: Face of the pulley is flat in the centre and tapers towards the ends forming a trapezoidal shape. They provide enhanced belt tracking capability as compared to taper crown while providing an even wear surface, lengthening belt life.

Crowned pulleys are most effective on conveyors with short, low-tension belts. With higher tension, steel-cable belts, and troughed conveyors, little steering effect is obtained from the crown of the pulley. Crowns are most effective where there is a long unsupported span (four times belt width or greater) approaching the pulley. Crowned pulleys should never be used for any pulleys on conveyors using steel-cable belt. Crowned pulleys should not be used for multi-ply belts at high tension; however, for all belts with textile carcasses, the best recommendation is that crowned pulleys be limited to locations where the belt will only be subjected to less than 40% of its rated tension.

6.1.4 Wing pulleys

Wing pulleys have a number of steel wing plates that extend radially from the longitudinal axis of two compression-type hub assemblies and are equally spaced about the pulley circumference. The purpose of compression-type hubs is to provide a clamp fit on the shaft. This construction results in the creation of open voids that are designed to allow loose material to fall away from the contact surface. Wing pulleys are primarily used on the tail end of the conveyor where loose materials have a tendency to reside on the underside of the conveyor belt, causing damage to one or both components.

Wing pulley used as tail pulley is often installed as a method to reduce the damage of material impingement. The wings allow the material, otherwise be trapped between a solid pulley and the belt, to pass through the face and fall to the ground without damage. Wing pulleys provide a self-cleaning function as well. There is a less surface for material to accumulate on the surface of wing pulley, and the spiral turning action tends to spin accumulations off the pulley's face. The spiral action introduces an oscillating action to the belt. The wings introduce a pulsating motion that makes noise and destabilizes the belt path, which adversely affects the belt sealing system.

The spiral wing pulley is another type of wing pulley used in applications that necessitate reducing material build-up between the belt and pulley while also utilizing the wing principle of self-cleaning pulley. In a spiral wing pulley, wrapping a band of steel in a spiral around the wing pulley allows the pulley to achieve continuous contact with the conveyor belt which eliminates excessive noise and pulsating motion without sacrificing the self-cleaning action. The cleaning action begins automatically when rotation begins and material is discharged to both sides of the conveyor. This design minimizes belt misalignment and decreases belt wear.

Wing pulleys should be avoided if at all possible because they cause the belt to vibrate transversely and vertically, making it difficult to clean the belt and seal the transfer point. Wing pulleys will shorten the life of the belt and the mechanical belt splice. Figure 6.5 shows two types of wing pulleys.

(a)   **(b)**

Figure 6.5 Wing pulleys. (a) Wing pulley and (b) spiral wing pulley. (Courtesy: ABB Motors and Mechanical Inc.)

6.1.5 Pulley lagging

Lagging is nothing but covering the surface with some other material. Conveyor pulleys can be lagged with some form of rubber, fabric, ceramic, belt strips, synthetic materials such as polyurethane, wood slats or other material. Lagging thickness varies from 0.8 to 50 mm. Generally, pulleys are provided with rubber lagging.

The advantages of lagging are as follows:

1. Rubber lagging on drive pulleys increases the coefficient of friction between the belt and pulley and maintains the friction grip even under unfavourable driving conditions such as those resulting from moisture and belt coating.
2. An increase in friction permits the use of lower initial belt tension.
3. Lagging reduces the wear on the pulley face, due to which the life of the drive pulleys increases.
4. Lagging effects the self-cleaning action on the surface of the pulley.
5. Lagging reduces the level of noise.

Lagging can be done by bolting, sliding, tack welding, painting, cementing and vulcanizing. Bolted-on and slide-on laggings have the advantage of being replaceable in the field. Welded lagging can be either a slide-on or weld-on lagging. Slide-on lagging works well with bidirectional pulley rotation. Replacement of slide-on and weld-on lagging can be done without removing the pulley from its conveyor location. Painted-on and cemented laggings are generally limited to lighter service conditions and are normally not used where grooving or embedment characteristics are needed.

Vulcanized lagging is generally preferred for heavy-duty or severe service applications. Vulcanized lagging can be done by hot process or cold process. Hot lagging process cannot be done at the site and has to be done under controlled conditions, with autoclave. Hence, the pulley has to be removed, dismantled and sent to the works. This would depend on the availability of spare pulley. Cold lagging can be done at the site.

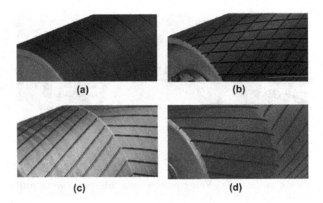

Figure 6.6 Common types of grooves on lagging. (a) Circumferential, (b) diamond, (c) chevron and (d) herringbone. (Courtesy: ABB Motors and Mechanical Inc.)

Pulleys can be lagged with plain rubber. They have a smooth surface and protect non-drive pulleys on the dirty side of the belt. Drive pulleys that perform in wet or damp conditions are often grooved. The common types of grooves are circumferential, diamond, chevron and herringbone (Figure 6.6).

Circumferential grooves are used on non-drive pulleys for self-cleaning. Diamond shape grooving is primarily used for reversing conveyor drive pulleys. It is also often used for spare pulleys when one does not know the direction of rotation. Diamond shape grooving sheds water from the belt and reduces spare pulley inventory. In a chevron pattern, the grooves meet at the pulley centre, while in the herringbone pattern, grooves are offset by one-half the groove spacing. In both patterns, the apex points in the direction of belt travel. The purpose is to improve traction between the belt and the pulley, thus lowering the belt wear and minimizing alignment problems.

There are also other groove configurations and sizes of rubber lagging which can be furnished to enhance belt tracking or to reduce the accumulation of material on the pulley face. Grooving of lagging does not increase the value of friction but only helps to squeeze out the water layer, establishing contact with the belt and pulleys. Grooving of lagging splits the film of water and deflects the water, so that the belt can rest on the pulleys. Whatever the type of lagging, it should be ensured that the lagging does not harm the belt. A good lagging reduces the belt tension, and hence, there will be saving on the cost of the belt.

With steel cord and other high-modulus conveyor belts, lagging is almost always used on drive pulleys and is preferred on pulleys that contact the carrying side of the belt. For maximum pulley life and belt safety, lagging can be used on all non-drive pulleys. Lagging for high-tension applications should be solid vulcanized rubber machined concentric with the shaft.

Wing pulleys are also lagged. A lagged wing pulley is used in cases involving build-up of material between the conveyor belt and pulley causing belt misalignment and excessive wear. It can also be used in cases requiring a greater tractive capacity between the belt and the pulley. The lagged wing pulley principle works as follows. Upon contact with the belt, the rubber on the wing tip compresses, and when tension is relieved, it immediately returns to its original shape, thereby cleaning itself in the process. These wing tip rubber sections are easily replaceable.

6.1.6 Ceramic lagging

Ceramic lagging is used in case of belt slip or high wear. Ceramic lagging commonly consists of a series of tiles embedded in a rubber substrate forming a bar profile. The tiles may be smooth or have raised surfaces on each tile. Those with raised surfaces tend to have better drive characteristics under wet, sloppy conditions. Due to the raised surface on the tile and the nature of ceramic, this type of lagging exhibits a superior coefficient of friction and greater wear resistance than rubber lagging. Depending upon application conditions, dimpled ceramic lagging can provide approximately two times higher traction than rubber lagging. Ceramic lagging also incorporates grooves, which are configured based upon the ceramic tile layout.

6.1.7 Pulleys for different purposes

The main purpose of pulleys in a belt conveyor is to run the belt by wrapping the belt around the pulleys arranged at both the ends. In addition to it, the pulleys also perform several functions, and they are named differently depending on their functions. Some of the pulleys are snub pulley, bend pulley, take-up pulley and magnetic head pulley.

Drive/Head Pulley: It is a pulley used for the purpose of driving a conveyor belt. Typically, it is mounted in external bearings and driven by an external drive source.

Return/Tail Pulley: It is a pulley used for the purpose of redirecting a conveyor belt back to the drive pulley. Tail pulley can utilize internal bearings or can be mounted in external bearings and is typically located at the end of the conveyor belt. Tail pulley also serves the purpose of a take-up pulley on conveyors of shorter lengths.

Snub Pulley: It is a pulley used to increase the arc of contact (wrap angle) between the belt and drive pulley, typically for the purpose of improving traction, and to relieve return rollers of excessive loading.

Bend Pulley: It is a pulley used whenever it is required to change the direction of belt from an inclined path to a horizontal path. It redirects the belt and provides belt tension where bends occur in the conveyor system. Bend pulley should be used when space will not permit a properly designed convex curve and the belt conveyor is not sufficiently loaded to cause spillage of material over the edges of the flattened belt as it passes over the bend pulley. Under these conditions, the diameter of the bend pulley should be large enough to ensure retention of the material on the belt as the belt changes direction.

Take-up Pulley: It is a pulley used to remove slackness and provide tension to a conveyor belt. Take-up pulleys are more common to conveyors of longer lengths.

Magnetic Head Pulley: It is a pulley used to remove the tramp iron. Magnetic head pulley holds the iron on the pulley past the material discharge point and then drops it down a separate pile or chute.

All types of pulleys installed on conveyor structure are shown in Figure 5.25.

6.1.8 Pulley diameter

The belt carcass is much stiffer as compared to rubber covers, and hence, carcass determines the pulley diameter. Belt carcass construction and tension in it decide the pulley diameter. The stiffer carcass in a belt needs a pulley of bigger diameter. The primary object in selecting adequate pulley diameters is to make sure that, as the belt wraps around the pulley under tension, the stress in the belt carcass is below the fatigue limit of the bond between the belt components. Over-stressing of the belt carcass will result in ply separation and premature belt failure, particularly at the belt splices. In some applications where available space is limited, smaller-than-recommended pulleys may be used. This, however, will result in more frequent splice replacement and will affect the service life of the belt.

The minimum pulley diameter recommended for a particular belt depends upon three factors:

1. Carcass thickness – the wire rope diameter in the case of steel cord belts
   – The overall thickness of all plies plus the rubber skims between plies in the case of ply-type belts
   – The overall thickness of the thick woven fabric separating the top and bottom covers in the case of solid-woven belts
2. Operating tension – the relationship of the operating tension of the belt at the particular pulley to the belt's allowable working tension
3. Carcass modulus – the relationship between elongation of the carcass and the resulting stress

Whatever the carcass type, steel cord, ply-type or solid-woven, when a belt is passing over the pulley, the carcass portion on outer radius stretches, whereas carcass portion on inner radius contracts. At a given tension, if the radius of the pulley is too small, the elastic limit of the outer carcass portion may be exceeded and fracture, and at the same time, the compression of the inner carcass portion may cause severe crinkling and eventual ply separation.

The belt under higher tension will have less room for putting additional bending stress, and therefore, it will need bigger pulley. The belt with lower tension has more room for putting additional bending stress and permits use of smaller diameter pulley. Thus, the same conveyor has different-diameter pulleys at head, tail, take-up, bend and snub due to different values of belt tension at these locations.

Many factors are involved in the selection of the proper pulley diameter such as the amount of wrap, belt tension at the pulley, available space, characteristics of the material handled, belt life expectancy, shaft and bearing size, and size and ratio of reducer. Large pulleys require more space, greater torque and larger speed reducers.

Since the elastic properties of the rubber or PVC cover material are so much greater than those of the carcass material, the cover thickness of the belt is not a factor in determining minimum pulley size and may be ignored. Belt conveyor manufacturers' handbooks indicate recommendations for minimum pulley diameters.

6.1.9 Pulley face width

Face width is defined as the length of rim, wing or contact bar along the shaft axis. As all belts tend to wander a bit in operation, the overall face width of the pulleys should exceed the belt width if serious edge damage is to be avoided. The width of standard pulleys exceeds belt width by 50 mm for belts up to 100 cm width. However, for belting width above 150 cm, the pulleys are 100 mm wider than the belt. Special conditions, such as very long conveyors, complex terminals or when handling sticky material, can dictate even greater dimensional differences.

6.2 Shafts

Shaft is a rotating element, normally cylindrical in cross section, and connects the pulley and motor for transmission of power from motor to pulley through reduction gear, which reduces the speed of the pulley to the required level. In Figure 6.7, (i), (iii), (v) and (vii) show different arrangements where pulley is connected to gear reducer and then to motor through shaft by direct coupling. Coupling also facilitates easy detachment for necessary maintenance and/or repairs of motor, gear reducer, pulley, etc.

The material used for ordinary shafting is mild steel. When high strength is required, an alloy steel such as nickel, nickel–chromium or chrome-vanadium steel is used. High-strength shafting is of value in cases where it permits the shaft ends to be turned down so that smaller diameter, high-capacity anti-friction bearings can be used.

It is essential that the diameter of the shaft, shaft material and bearing centres be known for design. Extreme caution must be exercised to avoid excessive shaft deflection, which can increase the stress and deflection in pulley end discs of the standard welded steel pulleys. A pulley made with very thick end discs has a minimum shaft deflection, and therefore, only the manufacturer can determine the actual deflection. Hence, it is required to obtain complete pulley loading information while procuring the pulleys.

6.3 Drives

All belt conveyor installations involve the proper application of conveyor drive equipment including speed reduction, electric motors and controls and safety devices. The preferred drive location for a belt conveyor is that which results in the least maximum belt tension. For simple horizontal and inclined conveyors, this is usually at the discharge end. For decline conveyors, the preferred location is usually at the load end. Special conditions and requirements can require that the drive be located elsewhere. Often internal drives are utilized on longer conveyors and inclined boom conveyors for reasons of economy, accessibility or maintenance.

Belt conveyor drive equipment normally consists of a motor, speed reducer, drive shaft and necessary machinery to transmit power from one item to another. The simplest arrangement of drive equipment using the least number of components is the best. Often, however, special-purpose components must be provided to modify starting and stopping, provide for a holdback or vary belt speed.

Figure 6.7 shows some of the more commonly used drive equipment assemblies.

i.   Gearmotor directly connected by flexible coupling to drive shaft.
        It is a simple, reliable and economical drive.
ii.   Gearmotor combined with chain drive to drive shaft.
        It is the one of the lowest-cost flexible arrangements and is substantially reliable.
iii. Parallel-shaft speed reducer directly coupled to the motor and to drive shaft.
        It is versatile, reliable and generally heavier in construction and easy to maintain.

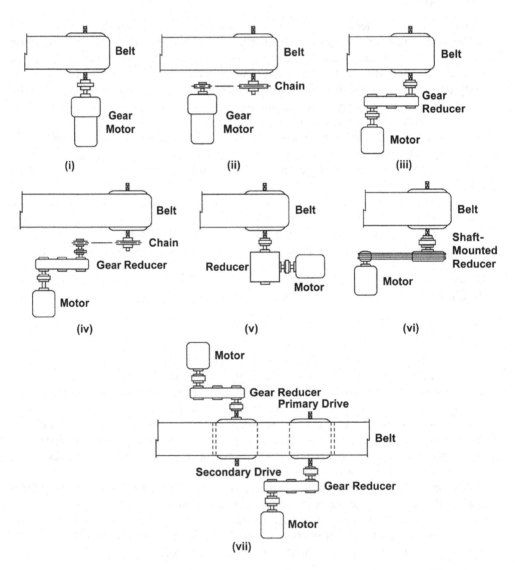

Figure 6.7 More commonly used drive equipment assemblies. (Courtesy: CEMA.)

iv. Parallel-shaft speed reducer coupled to motor and with chain drive, to drive shaft.
It provides flexibility of location and also is suitable for the higher horsepower requirements.

v. Spiral-bevel helical speed reducer, or worm-gear speed reducer, directly coupled to motor and to drive shaft.
It is often desirable for space saving and simplicity of supports. The spiral-bevel reducer costs substantially more than the worm-gear speed reducer but is considerably more efficient.

vi. Drive-shaft-mounted speed reducer with V-belt reduction from motor.
It provides low initial cost, flexibility of location, and the possibility of some speed variation and space savings where large speed reduction ratios are not required and where horsepower requirements are not too large.

vii. Dual-pulley drive.
It is used where power requirements are very large, and use of heavy drive equipment may be economical by reducing belt tensions.

The final selection of the speed reduction mechanism is based on preference, capital cost, power limitations, speed reduction characteristics, space and/or drive locations. There is also a motorized head pulley available, which combines the motor and reducer within the framework of a pulley. These units must be carefully selected initially because once a choice is made as to horsepower and speed, any change could require an entire unit replacement.

The general-purpose squirrel-cage motor will fulfil the requirements of most belt conveyor drives. Open motors are objectionable in dusty or exposed locations. Splash-proof or totally enclosed fan-cooled motors should be used whenever drives are not housed. High-torque motor is needed for conveyors requiring large amount of power. Wound-rotor motors with step starting, squirrel-cage induction motor with auto-transformer and eddy current coupling are the three electrical devices used to control the acceleration when conveyor is started. Smooth starting is accomplished by the use of torque couplings. Fluid coupling, variable-speed hydraulic coupling, dry fluid coupling and centrifugal clutch coupling are the few mechanical devices used for controlling the acceleration.

The most common mechanical methods of obtaining variable speeds of belt conveyors are V-belt drives on variable-pitch diameter sheaves or pulleys, variable-speed transmissions and variable-speed hydraulic couplings.

Backstop is used to prevent the reversal of motion of belt conveyor. Backstop is a mechanical device that allows the conveyor to operate only in the desired direction. It permits free rotation of the drive pulley in the forward direction but automatically prevents rotation of the drive pulley in the opposite direction. Three general backstop designs are ratchet and pawl, differential band brake and overrunning clutch. Brake is a friction device used for bringing a conveyor belt to a controlled stop and also used to control the coasting distance. Mechanical friction brake and eddy current brake are the two types of brakes used in belt conveyors.

Chapter 7

Belt take-ups, cleaners and accessories

7.1 Belt take-ups

Belt take-up is an assembly of structural and mechanical parts which provides the means to maintain the proper tension in the belt and to adjust the length of the belt to compensate for stretch. Take-up devices derive their name from the fact that they take up changes in belt length. In taking up length, they maintain tension, which is their primary purpose. Hence, every belt conveyor is equipped with a take-up.

While the conveyor is running, all tension variations are due to start-up acceleration, stopping/deceleration and change in the load. Some belts have multiple loading points and can have rapid changes in tension. The belt conveyor take-up device is a vital component in ensuring optimum belt performance by maintaining required tension in the belt during different conditions causing variations in tension.

All belt conveyors require the use of some form of take-up device for the following reasons:

1. To ensure adequate slack-side tension at the drive pulley to prevent belt slip
2. To ensure adequate belt tension at the loading point and at any other point of the conveyor to keep the troughed belt in shape to avoid material spillage and limit belt sag between carrying rollers
3. To compensate changes in belt length due to stretch
4. To allow belt storage for making replacement splices without having to add an extra piece of belt

Hence, a belt conveyor take-up has three functions:

1. It applies the initial tension necessary to provide tension between the driving pulley and the belt.
2. It removes the slackness in the belt due to belt stretch.
3. It provides sufficient reserve belt length, to enable a new joint to be made, if necessary.

Two types of take-ups, screw take-up and gravity take-up, are generally used:

1. **Screw Take-up** (fixed type, manually adjusted)
   The screw take-up (Figure 7.1) is most commonly used manual take-up. In a screw take-up system, the take-up pulley (usually tail pulley) rotates in two bearing

Figure 7.1 Screw take-up.

blocks that may slide on stationery guideways with the help of two screws. The tension is created by the two screws that are manually tightened and periodically adjusted with a spanner. Manual take-ups have the advantages of compactness and low cost.

The main problem with the use of a manual take-up is that it requires a vigilant and careful operator to observe when take-up is required, and then adjust the take-up just to the point where the proper tension is provided. The operator must rely on personal judgement in making the suitable adjustment.

The screw take-up arrangement is generally installed with the tail pulley, which also acts as the take-up pulley. This is the most convenient and least costly arrangement as it involves no extra pulleys. The screw take-up is normally used on short conveyors up to a length of 60 m and light duty. Over 60 m, an automatic take-up system should be used to ensure constant pre-tension. The length of the take-up depends on both the centre distance of the conveyor and the type of the belt. Tracking of the belt with the screw take-up should only be performed as a last resort as it may cause permanent belt stretch.

2. **Gravity Take-up** (counterweighted take-up, floating type)

In gravity take-up, the pulley positioning device applies a constant tension to the belt and the pulley location is determined by the belt strength and other factors. In gravity take-up devices, take-up pulley is mounted on slides or on a trolley and travels freely while a constant tension is automatically maintained to ensure normal conveyor operation in all cases. In these devices, because pulley shifts down/up/sides in relation to belt stretch/shrinkage due to variation in tension, temperature, etc., they are also called floating take-up pulley type devices. Since they automatically compensate for belt stretch under varying operating conditions, they provide the lowest tension around the belt path.

Gravity take-up devices may be located at any point along the return run or at the tail pulley. These are usually preferred for long-centre conveyors and have the following features:

- It is self-adjusting and automatic.
- Greater take-up movement is possible.
- It is suitable for horizontal or vertical installation.
- It can be located at drive end (preferred for low tensions).

In gravity take-up, there are two types:

1.   Carriage take-up
2.   Vertical take-up

In a carriage take-up, also called as horizontal take-up, as shown in Figure 7.2, take-up is provided by mounting the tail pulley on a trolley/carriage that runs on tracks (horizontal guide frame) parallel to the centre line of the belt. A system of wire rope and deflecting pulleys to a suspended counterweight at any convenient location keeps this trolley pulled back and establishes belt tension.

The take-up location at tail end is economical because it does not require extra pulleys and extra length of the belt. However, if the conveyor is very long, it will have delayed response to the drive, due to large inertia of the return run. This can cause very small momentary slippage between the belt and the drive pulley, resulting in faster wear of belt and drive pulley periphery. In such cases, the take-up location at intermediate point can be chosen closer to the drive, resulting in quicker response to the needs of the drive. This will be due to less inertia of moving mass between take-up and the drive.

The vertical take-up as shown in Figure 7.3 consists of three pulleys, two deflecting pulleys and one tensioning (take-up) pulley, and is installed on the return strand of the conveyor. The tensioning pulley travels, together with the moving frame, along vertical guides and takes up belt slack by its own weight and that of the counterweight suspended to the moving frame.

The carriage-type take-up is superior to the vertical type because it is of much simpler design and of considerably less height. It does not, however, ensure uniform tension in the slack side of the belt. Losses on the rollers cause certain variations in the belt tension. A vertical take-up does not possess these drawbacks when installed near the drive pulley. The latter type is also better adapted for long take-up travel required to eliminate slack occasioned by belt stretch under wear and sag between the rollers. The three pulleys required

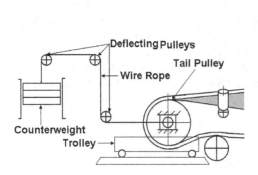

Figure 7.2 Carriage take-up.

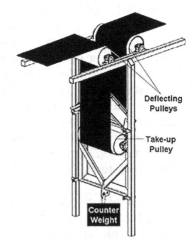

Figure 7.3 Vertical take-up.

and bends of the belt in different directions are drawbacks of the vertical arrangement.

The other two important types of take-ups are as follows:

1. **Winch Take-up**

    The winch take-up system is similar to the screw take-up system except longer take-up travel is more conveniently provided. A winch is a mechanical device used to pull in or pull out or otherwise adjust the tension of a rope. In its simplest form, it consists of a spool and attached hand crank. A steel rope, wrapped around a winch, exerts the pull on the take-up pulley. In some cases, the take-up pulley is moved by a gear and track arrangement. Winch take-up device can also be used as automatic take-up arrangement when automatic tension regulation by employing load cells, electronic sensing devices, etc., is provided to signal for the winch motor to run in one direction or reverse for a specific number of turns or to stop as governed by predetermined values of belt tensions for any particular installation.

2. **Hydraulic Take-up**

    The movement of the take-up pulley in this case is controlled by hydraulic cylinders directly or by hydraulic cylinders through steel wire ropes. In most hydraulic take-ups, pulley location is controlled by a hydraulic servo system.

    Screw take-ups are employed primarily in short horizontal and inclined conveyors where the ambient temperature and humidity are comparatively stable. Carriage-type arrangements are used preferably in conveyors of medium length (50 to 100 m) having a complicated path of motion. Vertical arrangements are used in long conveyors (100 m and over) and where lack of space precludes the installation of a carriage at the conveyor tail end. In long heavy-duty conveyors, carriage-type take-ups with automatic (or semi-automatic) belt tensioning devices are employed. These devices utilize a mechanical winch driven by a reversible electric motor. The tension of the slack-side belt is set to maximum when the belt is started and reaches a minimum when the conveyor runs steadily. This arrangement makes the belt tension a function of the working load.

7.2 Belt cleaners

Many bulk materials conveyed on belts are somewhat sticky. Such sticky material will adhere to the top surface of the belt conveying the material and will not discharge with the rest of the material at the discharge point. The residual material is carried back on the return run. This carryback material is dislodged from the belt surface by the vibration of the rollers and falls off the belt at various points along the return run of the belt.

The carryback particles are smaller in size with higher moisture content. These fines will settle to the bottom of the belt through the pile due to vibration caused as the belt rolls over the rollers. The fines with moisture form the adhesive mixture and cling to the belt. When it is removed, it is attached to other surfaces. These accumulations become hard when allowed to dry. The carryback material causes the following:

- Excessive wear
- Material build-up on return rollers resulting in misalignment of the belt and increased power consumption
- Damage to the belt by forcing the belt against some part of the supporting structure

- Reduced operating efficiency and profitability through increased expenses for maintenance and clean-up and the loss of material
- Unsafe working conditions caused by material accumulation on floors and walkways, creating fire hazards and slip/trip hazards
- Health hazards and environmental concerns created by airborne material
- Unfavourable attention from neighbours and regulatory agencies

Therefore, it is desirable to clean the belt before the carryback material clings to the snub pulleys. Minimizing the carryback material with the belt cleaners results in the following:

- Improved maintenance planning and conveyor availability as emergency outages.
- Reduced unscheduled downtime and 'hurry-up' repairs.
- Reduced maintenance expenses by lower labour costs for fewer and faster service procedures.
- Improved manpower utilization by fewer belt tracking and material clean-up chores.
- Maximized equipment life by fewer replacements of prematurely worn components damaged by fugitive material and build-up.
- Improved working conditions and plant safety and morale by better housekeeping.
- Improved community relations and regulatory compliance by reducing environmental pollution.

Conveyor belt cleaners are some form of scraper or wiper device mounted at or near the discharge pulley to remove the residual fines that adhere to the belt as it passes around the head pulley. It is usually best to install a multiple cleaning system composed of a primary cleaner and one or more secondary cleaners as shown in Figure 7.4.

The primary cleaner functions as a metering blade and removes the majority of material adhered to the belt, leaving only a thin layer of sticky fines. Primary cleaner, sometimes called as doctor blade, is typically installed on the face of the head pulley, just below the material trajectory. The cleaner should be constructed to avoid material build-up and installed so that it is out of the material stream.

Figure 7.4 Multiple cleaning system.

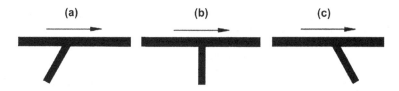

Figure 7.5 Angles of cleaning. (a) Scraping angle, (b) normal angle and (c) peeling angle.

Secondary cleaner is installed at the point where the belt is leaving the discharge pulley. Secondary cleaners remove residual fines that remain on the belt past the primary cleaner. Its location is typically close enough to the material trajectory that the cleanings will return to the main material stream. An additional tertiary cleaner can be installed to provide final cleaning. This cleaner can be the same model as the secondary cleaner, or of a different design to allow efficient cleaning and maintenance within the available space. As tertiary cleaners are typically installed away from the pulley, they should be placed at or near a point where the belt is against a roller.

The angle of attack of the blade to the belt is an important consideration. Three angles, viz. **scraping angle**, **normal angle** and **peeling angle**, are shown in Figure 7.5.

In the case of scraping angle position, the blades are inclined with the direction of belt travel. In this position, the carryback material builds up in the edge and develops a high pressure leading to high belt wear.

Normal angle position is the oldest system. In this position, the rubber blade, which may be installed on or after the head pulley, is perpendicular to the surface. This is pressed against the belt by means of a counterweight or spring action. In this, a high-pressure turbulent zone is formed, leading to rapid wear on its leading edge.

In the case of peeling angle position, the blades are fixed in such a way that they are inclined against the direction of belt travel. In this position, no build-up of material is likely to occur. High-pressure zone is minimum, and hence, lower pressure is exerted on the belt. Further, as the blades wear down, they generate a sharp leading edge, which is maintained for the life of the blade, ensuring effective cleaning.

In a multiple cleaning system, the primary cleaner's edge is positioned against the belt in a peeling position with a low angle of attack. This allows a non-metallic blade to clean effectively by shearing the majority of loosely adhered material away from the belt with the application of minimal pressure. The secondary cleaner is designed to remove the remainder of the material, the sticky fines which have passed under the primary cleaner's blade. The secondary cleaner is to be positioned in contact with the belt as it leaves the head pulley and in the direction of the belt's travel in a scraping angle.

7.2.1 The ideal scraper design

The ideal scraper must operate on the face of the discharge pulley to deposit the material in the discharge area. At this position, fewer variables are encountered and the results obtained are more consistent. A scraper should have a constant geometry so that the forces do not alter the behaviour of the scraper when they change.

An ideal belt cleaner should conform to the following criteria. It should:

- Operate on the face of the discharge pulley
- Be contained within the chute
- Be reversible
- Withstand high temperatures
- Withstand corrosion
- Be resilient to obstructions
- Be safe for the belt
- Have a large number of small blades
- Ensure contact of the blades at all points across the dirty section of the belt
- Not require maintenance at short intervals
- Be very simple to maintain
- Be efficient at any belt speed
- Be efficient at any moisture content

7.2.2 Types of belt cleaners

Basically, the cleaners (or scrapers or wipers) can be divided into two types as **static type** and **dynamic type**.

7.2.2.1 Static-type cleaners

These are generally suited for the carrying and return sides of the belt. A few devices are

Manila, hemp or coir rope
Piano wire
Screw rollers
Simple wiper
Double wiper
Tail protection ploughs (V-plough and straight plough)
Segmented blade

Figure 7.6 shows the simplest form of belt wiper. Manila, hemp or coir rope of 20/25 mm diameter is stretched across the belt with two clamps. The rope makes light contact with the belt. Depending on the amount of wear, the rope has to be replaced. For light-duty conditions, such as cement or sand transport, this is effective and does not harm the belt.

Figure 7.7 shows a mechanism where a piano wire is used. A piano wire of about 5 mm diameter is tensioned across the belt at the head pulley with 0.5 mm clearance to the vulcanized belt. The results have shown that a less than 0.5 mm film is left on the belt to be removed by a standard blade-type device.

Return spiral rollers and disc rollers, as shown in Figures 5.13 and 5.14, can be used effectively for cleaning the return side of the belt. It should be ensured that these are replaced in time. Otherwise, there is a possibility of belt damage due to thinning out of the rods or shells.

Figure 7.8 shows a simple wiper or cleaner cleaning the belt. Figure 7.9 shows a double-wiper system where the first wiper or primary cleaner does the major cleaning, whereas the second wiper or secondary cleaner cleans the balance material.

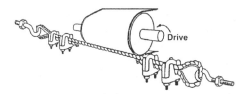

Figure 7.6 Rope-type belt wiper.

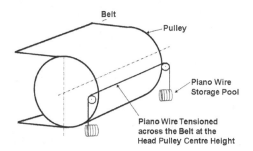

Figure 7.7 Piano wire device.

Figure 7.8 Simple wiper system.

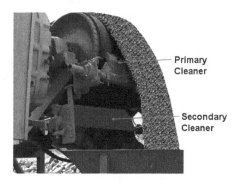

Figure 7.9 Double-wiper system.

Plough wipers are used for cleaning the surface of the return belt. They clean the belt and prevent build-up of material and dust on the tail pulley, thus keeping the belt properly aligned. Figures 7.10 and 7.11 show V-plough and straight plough. Originally, the plough wiper was meant to prevent damage to the belt by inadvertent falling off of troughing rollers or carrying rollers from brackets. These fallen rollers are likely to be carried by the return strand of the belt to the tail end pulley, where due to jamming of the roller against the pulley, the belt is likely to get damaged. The plough wiper prevents such incidents by pushing the fallen roller away from the belt.

Figure 7.10 V-plough.

Figure 7.11 Straight plough.

Plough wiper is positioned on top of the return strand near the entrance to the tail pulley. Plough wipers of V configuration are usually constructed with steel frame/ rubber blade. On reversible belts, a straight plough with a single blade is normally installed diagonally across the belt, ready to deflect material when the belt is running in either direction.

Figure 7.12 shows a segmented blade-type wiper. Segmented blade types include both single- and double-row designs. Their main advantage is that they can cope with a curved belt profile and offer an increased area of contact, due to the overlap of adjacent segments. Material build-up on this type of device can be problem in severe, wet and/or sticky conditions. If this situation arises, a single-blade device should be fixed prior to the segmented blade.

Use of Water Spray: Depending on the material carried, water spray is one form of effective cleaning of the belt, combined with mechanical scrapers. Apart from cleaning efficiency, water spray reduces the belt and blade wear. A few disadvantages of this system are as follows:

- Availability of continuous water supply
- Control of the correct quantity of water
- Effective stoppage of water spray when the conveyor is not operational
- Difficulty in keeping the area clean
- Safety hazard, if cleaned materials are not cleared immediately
- Difficulty in protecting nearby electrical equipment, etc.

Use of Air Stream: Air stream is also used for cleaning the belt. Air stream is directed over the face of the pulley to shear off the carryback material. It is effective to remove dry or extremely wet materials. For certain applications, vacuum could be used for cleaning the belts.

Belt Turnover Device: Belt turnover device is a device used for keeping the dirty side up on the return side for cleaning purposes. To eliminate the problems caused by a dirty belt in contact with return idlers, the belt can be twisted 180° after it passes the discharge point. This brings the clean surface of the belt in contact with the return idler rolls. The belt must be turned back again 180° before entering the tail section, to bring the conveying side of the belt up at the loading point. Abnormal tensions are induced in the carcass of the belt when the belt is turned. Hence, few manufacturers

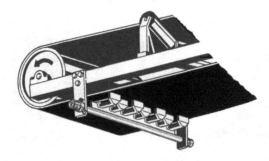

Figure 7.12 Segmented blade-type wiper.

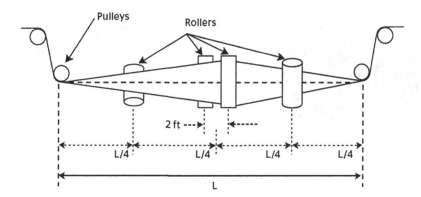

Figure 7.13 Belt turnover.

have technical objections to this method. The distance required to accomplish the 180° turn of the belt is approximately 12 times the belt width. Short transition distance can cause the belt to collapse in the turnover or cause excessive edge tension, which can in turn cause premature splice failure. This belt turning method is useful on very long conveyors.

The turnover must have two terminal pulleys – inbound and outbound – as shown in Figure 7.13. This is to ensure that the belt is lying flat on the first bend pulley entering the transition and the last bend pulley leaving the transition.

Vertical rolls should be located near the mid-point and 45° rolls located at quarter locations of turnover. The mid-point rolls should be on opposite sides of the belt and offset a few inches from each other so that a component of belt tension can provide pressure against each. They should be adjustable in all directions. These rolls aid in belt training, help minimize belt buckling tendencies and help stabilize a belt from excess flopping where winds are encountered.

Make sure the spacing between the vertical rolls is 2 ft (Figure 7.13). Each vertical roll also needs to be offset by 1″ inward so that there is full belt contact on both rolls. Each end pulley should have some adjustment in the plane of the approaching belt for training purposes.

7.2.2.2 Dynamic-type cleaners

Dynamic cleaning devices comprise two distinct groups:

1. **motor-driven**
2. **belt-propelled**

Motor-driven devices usually involve rotating drum units, with the cleaning effect being carried out by rubber or plastic fingers, spirals or bristles. Rotary brush belt cleaner (Figure 7.14) consists of rotary brushes, which have a flicking action, usually having bristles in a helical or parallel pattern designed for fitment below the head pulley on the return side to enable the brushed material to fall into the discharge chute.

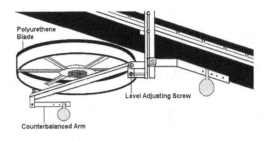

Figure 7.15 Rotary scraper – belt driven.

Figure 7.14 Rotary brush belt cleaner.

It consists of a motorized power-driven shaft to which brush rolls are fitted as shown in Figure 7.14. The direction of rotation is opposite to that of the return belt travel. V-belt drive can also be used for brush rolls. The V-belt drive ensures that right peripheral speed of the rotary scraper can be achieved to get maximum cleaning efficiency. Low-speed and high-speed rotary brush cleaners are the two types of rotary brush belt cleaners.

Rotary blade belt cleaners use rubber blades arranged parallel to, or helically on, the shaft. These rubber blades have a squeegee or scraping action. Low-speed and high-speed rotary blade belt cleaners are the two types usually made.

Rotary scraper is the belt-propelled cleaner. The rotary motion to rotating wiper is imparted due to the belt movement itself. The rotating unit consists of a base plate with two cross-channels which has got four fixtures for fixing the same at different heights. The base plate is a ring-shaped cleaning mechanism that is mounted on a central bearing. The wiper should be fixed slightly at an angle to the running direction of the belt so that sufficient pressure is exerted by the belt to be cleaned. This will help the continuous rotation of the ring-shaped wiper, which effects efficient cleaning (Figure 7.15).

Dynamic devices are best incorporated at the design stage, so that they can be positioned within the discharge chutes. Generally, it is very difficult to install these units into the head chutes of existing conveyors, and if installed as an addition, they have usually to be fitted in a secondary position.

Care has to be taken to ensure that motor-driven dynamic devices are adequately housed to contain the scattered cleanings and protect the motor from contamination. To be more effective and have a minimal maintenance requirement over a reasonable period, dynamic devices also benefit from an automatic pressure-and-wear adjustment system. In general, dynamic devices are more expensive to purchase, install and operate than static types. However, this does not automatically mean that in the long term, they are more costly. This point needs careful consideration when choosing between static and dynamic types.

The problem of keeping a belt clean is a difficult task unless this is done systematically and regularly. The cleaning system should be inbuilt into the system, from the very beginning. Belt being the costliest component in a conveyor and the amount and time spent in achieving an effective cleaning system will be paid multifold, by the increase in the life of the belt, and reduced spillage clearance.

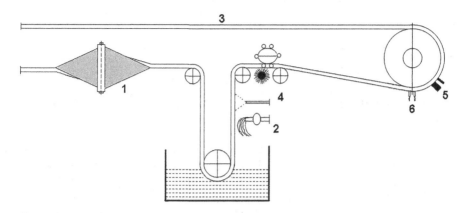

Figure 7.16 Main categories of belt cleaning procedures.

A common method of cleaning the pulley surfaces is to mount a bar scraper on the ascending side of the pulley, about 45° below the horizontal centre line, so that the material scraped from the pulley will fall free. The scraper bar should have a hardened bevel scraping edge. Also, it should be adjustably fixed in position on a stiffener so that the bevelled edge of the scraper blade just clears the pulley face and works like a chisel to shave off material that clings to the pulley.

7.2.3 The present state of belt cleaning technology

The conveyor belt cleaning devices currently available on the market can be broadly divided into the following categories:

- 1 - Belt turning device for the return run
- 2 - Belt washing, spraying and air jetting equipment
- 3 - Treatment of the belt faces, or the top wearing layer of the rubber cover, so as to reduce adhesion
- 4 - Devices for rapping, brushing, wiping or frictional stripping for the removal of adhering material with the aid of roller, brushes or rotors
- 5 - Belt scraping devices with continuous linear contact by means of bars, beams or wires
- 6 - Belt scraping devices based on intermittent linear contact by means of individually adjustable and, in some instances, spring-loaded scraper blades, usually of metallic construction

The main categories of belt cleaning procedures are shown in Figure 7.16.

7.3 Accessories

In addition to the major components of the belt conveyor system, a number of accessories, such as belt conveyor protection devices, tramp iron detectors and removers, conveyor belt scales and sampling devices, are also required.

7.3.1 Protection of belt conveyor

Belt conveyors are to be protected from the following:

- weather
- volatile characteristics of the conveyed material
- plant works
- various standards that require the covering of the total length of a belt conveyor in the open

Protection of belt conveyor from the weather varies with the climate, the material handled and the type of operation. The material handled may be affected by rain or, if the material contains moisture, by freezing temperatures. Rain on the pulleys may create slippage of the belt, causing a tracking problem. Ice and snow on pulleys can completely stop a belt conveyor, if the drive cannot move the belt. Intense sunlight may shorten the life of rubber covers on the belt. Very strong winds may lift conveyor belts off their rollers, blow fine materials from the conveyor belt and cause serious damage to the belt. Extreme temperatures may cause the plant to malfunction or stop.

Galleries and **housings** are used to enclose belt conveyors. Galleries are structures that combine the structural support of the conveyor and protection from the weather while providing enclosure to reduce the release of fugitive material and make cleaning easier. A gallery is typically a structural steel frame covered with metal or plastic siding. Galleries offer most complete protection for the conveyor parts and maintenance personnel. Housings prevent ice and wind from causing a belt to run off-centre and an empty belt from being blown off the rollers. They also decrease deterioration of the belt by protecting it from the sun.

Belt conveyor **covers** or **hoods** are used for belt protection when conveyors are not contained within an enclosed gallery. The cover or hood is an arc of rigid sheeting, held in place by hoop-like band. Conveyor covers are manufactured with smooth or corrugated galvanized steel, aluminium, stainless steel, fibreglass, painted steel and plastics. Three types of commonly used covers are full covers, three-quarter covers and half covers. The covers are used outside the loading and discharge chutes. They protect the belt, rollers and material conveyed from the sun, wind, snow, rain and other uncertain trash weather elements. They effectively improve the personnel safety, reduce loss of material to wind and stop grime and rain from damaging the rollers and belt.

When protection is needed from wind alone, **windbreaks** are used. Windbreak is a reinforced metal sheet on the windward side of the conveyor stringers and installed in such a way that the sheet extends above and below the stringers to protect both the carrying and return runs of the belt. **Wind hoops** are the round bar hoops spaced at regular intervals along the belt conveyor to prevent the empty or lightly loaded belt from being blown off the rollers.

7.3.2 Tramp iron detectors

Removal of tramp iron by arranging the magnetic pulley at the discharge end of the belt conveyor is one method commonly used. Suspended stationary magnet over the belt alone or in combination with a magnetic pulley is another method for removing tramp iron.

A third method is to arrange a short and cross belt conveyor with a stationary magnet. Tramp iron is picked up by the magnet arranged to the cross belt conveyor and held against the underside of the cross belt. As soon as this cross belt is moved from the magnetic field, the tramp iron drops into a container. The fourth method of removing tramp iron is to pass the conveyor belt through a magnetic field of adjustable intensity.

7.3.3 Conveyor belt scales

Belt scale is a device used for continuous weighing of material on a moving conveyor belt. The belt scale consists of three components, namely a weigh frame supported on load cells, a belt speed sensor and an electronic integrator with a PLC or microprocessor. The weigh frame with a load cell is inserted into a section of the conveyor. The load cells measure the material weight on the belt and send a signal to the integrator. The integrator also receives input in the form of electrical pulses from a belt speed sensor connected to a tail or bend pulley. Using these two sources of data, the integrator calculates the rate of material transferred along the belt using the following equation: rate = weight × speed.

All-mechanical system, electronic scale, laser scale and optical scale are the few types of belt scales.

7.3.4 Automatic sampling

Sampling is a method of collecting a small portion of a material which is representative to that of the whole material. The most suitable location for automatic sample collection from moving streams is the discharge end of belt conveyor. Here, the complete material stream is intercepted at regular intervals perpendicular to the material flow. These may be termed as belt end automatic mechanical samplers. This sampler is installed at the end of the head pulley. A cutter in this sampler traverses the stream and comes out of the stream, during which period the sample is collected from the entire cross section of a continuously moving stream. The cutter is driven across the stream by means of a hydraulic cylinder, electric motor or pneumatic drive.

Another method of automatic sampling is by using cross belt sampler consists of a scoop rotating perpendicular to the material flow above a conveyor. This sampler is set up directly above the belt conveyor. A thick steel scoop rotates and passes just above the belt. As it moves, it catches the slice of the material that is lying on the conveyor and pushes it towards a sample collection and gravity discharge point that is located at the edge of the conveyor.

Chapter 8

Transfer points

A transfer point is the point on a conveyor where a material is fed onto that conveyor or discharged from that conveyor. The successful operation of a belt conveyor requires proper feeding of the material onto the conveyor belt and proper discharge of the material from the conveyor belt. These two are the most important requirements for the functioning of the belt conveyor as desired.

A typical transfer feeding point (Figure 8.1) is basically composed of a metal chute to guide the flow of material while feeding the material from the storage vessel onto the conveyor belt. Two plates called skirtboards are arranged on both sides of the belt from the edge of the chute to some distance required to shape the material load on the belt and keep the material on the belt without spilling over its edges.

It may also include gate or feeder to regulate the flow of the material discharged from the hopper of storage vessel and a baffle or deflector within the chute to properly place the material onto the conveyor belt.

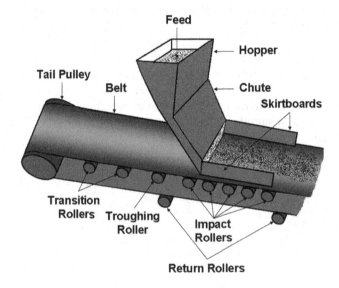

Figure 8.1 Transfer feeding point.

The transfer feeding point of any belt conveyor is the critical point and the life-determining point of the belt. It has an effect on spillage of the material and belt training. Conveyor belt receives its major abrasion and all of its impact at transfer feeding point. Since little can be controlled after the material has come to rest on the belt, the means of feeding the belt is a very important consideration in belt conveyor design. The success of a belt conveyor system greatly depends on the success of its transfer feeding point. If the material is fed poorly, conveyor belt, rolling components, and/or structure suffers damage decreasing its operating efficiency. The spilled-over material cause numerous maintenance problems leading to reduced production efficiency and increased operating and maintenance costs.

An ideal transfer feeding point should be designed to

- Feed the material onto the belt at a uniform rate
- Feed the material onto the belt centrally
- Feed the belt after the belt is fully troughed
- Reduce the impact of the material falling on the belt
- Deliver the material in the direction of belt travel
- Deliver the material to the belt at a velocity as near the speed of the belt as possible
- Maintain a minimum angle of inclination of the belt at the feeding point
- Provide adequate belt edge space to allow sealing
- Provide adequate space and systems to remove material carried back and feed it back
- Provide adequate space and access to allow inspection and service on both sides of the conveyor.

A transfer feeding point mainly is of two types:

1. It consists of a storage vessel with a gate or feeder at its outlet as shown in Figure 8.2 whenever the material is to be fed to the conveyor belt from a storage vessel.

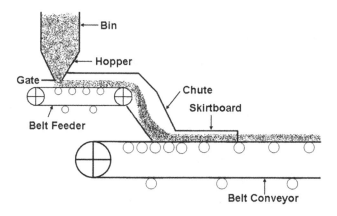

Figure 8.2 Material feeding from bin through belt feeder.

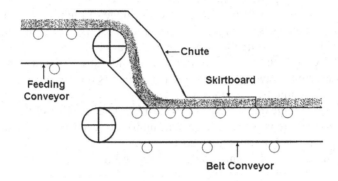

Figure 8.3 Material feeding from another conveyor belt.

2. It consists of a system of metal chutes to guide the flow of material as shown in Figure 8.3 whenever the material is to be fed to the conveyor belt discharged from another conveyor belt.

8.1 Material storage vessels

Bin, silo and bunker are the storage vessels used for storing bulk materials by every industry around the world. These three storage vessels often used to refer to similar vessels in different industries. Bin (Figure 8.4) consists of two main sections: cylinder and hopper. The cylinder section has a constant cross-sectional area over its height and is usually square, circular or rectangular in plan view. In the hopper, the cross-sectional area changes from top to bottom, usually decreasing (converging hopper). At the outlet, a gate, discharger or feeder is often used to stop and start the flow and, in many cases, to control the discharge rate. A silo is a larger size vessel than a bin. In a coal-fired power plant, the storage vessel is often called a bunker.

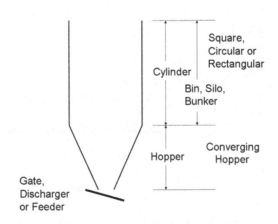

Figure 8.4 Bin.

An adjustable gate installed at the bottom of the hopper will allow the direction of the material flow path into the centre of the receiving belt. The gate should be adjustable from outside to make easy correction of flow of material. An automatic system of devices such as belt alignment switch or limit switch can be installed to control the placement of the movable gate.

Regular feeding of the material to the belt is very much essential as irregular feeding makes the belt empty and overloaded alternately, which leads to loss of capacity and spillage of material over the edges of the belt along the overloaded portions. In addition, a controlled and measured rate of material is required to be conveyed by the belt in plants where the material is processed at a definite rate. To achieve the regulated rate of material to be discharged from the hopper, a mechanical device, called feeder, is installed at the outlet of the hopper as shown in Figure 8.2 where a belt feeder is installed at the outlet of the hopper.

8.2 Feeders

Feeders are the mechanical devices intended to regulate the rate of flow of material. When once set, they should deliver the material at a definite rate as nearly as possible and they should be adjustable to feed fast or slow.

Feeders for controlling the flow of materials onto the conveyor belts require the following criteria to be met:

- They should deliver the range of flow rates required.
- They should handle the range of particles or lump sizes and flow properties expected.
- They should deliver a stable flow rate for the given equipment setting. They have to permit the flow rate to be varied easily over the required range without affecting the performance of the storage units from which they are feeding.
- They should feed material onto the conveyor belt in the correct direction at the correct speed with the correct loading characteristics and under conditions which produce minimum impact, wear and product degradation. Often, a feed chute is used in conjunction with the feeder to achieve these objectives.
- They have to fit into the space available.

Many types of feeders have been evolved to suit differing needs of material-handling and process plants. Belt feeders, apron feeders, screw feeders, vibrating feeders, reciprocating plate feeders, rotary drum feeders, rotary vane feeders, rotary table feeders, rotary plough feeders, drag-scraper feeders are the few feeders used in belt conveyor installation to regulate the feed to the belt conveyor. The choice of feeder depends upon the characteristics of the material handled, the manner of storage of the material and the rate of feed required.

8.2.1 Belt feeders

Belt feeders are one of the most widely used feeders. Belt feeders are very short belt conveyors, consist essentially of a continuous rubber or polymer belt running between end pulleys and are supported on a number of rollers. In normal use, they are fitted beneath a hopper as shown in Figure 8.2. Belt feeders are used for handling fine, free-flowing, abrasive and friable materials.

8.2.2 Apron feeders

Apron feeders (Figure 8.5), a version of belt feeders, consist of pans linked together and mounted on or between strands of conveyor chain fitted with rollers riding on central rail. Apron feeders are used for dealing with bulk materials which are very hard, abrasive and tough and for lumps of larger dimensions which are beyond the scope of belt feeders. They are useful for feeding large tonnages of materials and for those materials which require feeding at elevated temperatures. They are also able to sustain extreme impact loading.

8.2.3 Screw feeders

Screw feeders (Figure 8.6) consist essentially of a helical screw, driven by an external source and mounted beneath the hopper outlet. The main requirement for screw feeders is that there should be a uniform removal of material from the hopper outlet, and in this respect, screws with increasing pitch and increasing diameter are likely to be the most successful. Screw feeders are suitable for materials which are granular/powdery or which have small-sized lumps. Screw feeders are widely used for materials of low or zero cohesion such as fine and granular materials which have to be dispensed under controlled conditions at low flow rates. They are not suitable for the materials containing large lumps or highly aerated fines, or for those which tend to pack.

8.2.4 Vibrating feeders

Vibrating feeders (Figure 8.7) consist of steel pan or feed chute, which is freely supported or suspended and is vibrated electromagnetically in a direction oblique to its surface. The material in the pan is moved along the pan surface by this type of vibration. Its rate of movement depends on the amplitude and frequency of the vibration. Normally, a vibrating feeder is positioned under the opening in the bottom of a hopper or a hole under a storage pile. Vibrating feeders are useful for handling a wide range of materials. They are not suitable for fine materials of size −150 mesh and sticky cohesive materials.

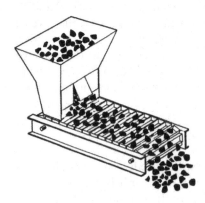

Figure 8.5 Apron feeder.

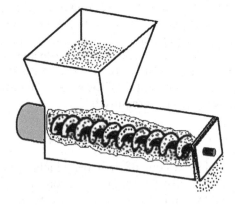

Figure 8.6 Screw feeder.

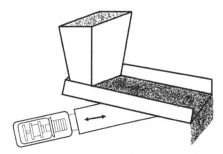

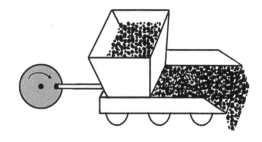

Figure 8.8 Reciprocating plate feeder.

Figure 8.7 Vibrating feeder.

8.2.5 *Reciprocating plate feeders*

Reciprocating plate feeders (Figure 8.8) consist of a reciprocating plate or pan. The reciprocating motion is imparted by crank or eccentric and connecting rod. In moving forward, a layer of material is carried forward on the plate and more material settles from the hopper onto the rear portion of the plate. On the reverse stroke, the plate slides under the material and that material at the forward end of the plate has its support withdrawn and falls. The plate is supported on wheels as shown in Figure 8.8 or suspended by rods. These feeders can handle fines, a mixture of lumps and fines, or small-sized lumps.

8.2.6 *Rotary drum feeders*

Rotary drum feeders (Figure 8.9) are simple and sturdy feeders and consist of a drum revolving slowly about a horizontal axis. The drum is located beneath the discharge of a hopper containing the material. By varying the speed of the drum or the width of the opening between the face of the drum and the discharge gate, the rate of feed may be varied. Rotary drum feeders are suitable for free-flowing fine and small-sized lump materials.

8.2.7 *Rotary vane feeders*

Rotary vane feeders (Figure 8.10), also called pocket feeders, consist of a shaft-mounted rotor vanes rotating in close clearance to its casing. As the vanes rotate, the pockets, which are formed between the vanes, become rotating pockets. The material enters the pockets at the top, through the inlet port, travels around in a rotating motion and exits at the bottom, through the outlet port. They are particularly used to discharge fine free-flowing materials from the storage units, while maintaining sealing so that air/gases do not flow into storage unit when the storage unit is under negative air pressure.

8.2.8 *Rotary table feeders*

Rotary table feeders (Figure 8.11), also called disc feeders, consist of a power-driven circular plate rotating directly below the storage hopper opening, combined with an adjustable feed collar which determines the volume of bulk material to be delivered.

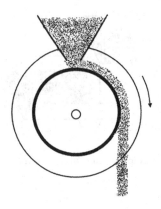
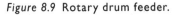

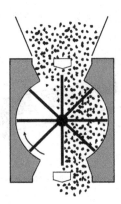

Figure 8.9 Rotary drum feeder.   Figure 8.10 Rotary vane feeder.

This permits equal quantities of material to flow from the storage hopper outlet and to spread out evenly over the table as it revolves. The material is then ploughed off by an adjustable blade and falls over the edge of the table or disc in a steady stream into a discharge chute. These feeders are suitable for non-abrasive and marginally abrasive materials.

8.2.9 Rotary plough feeders

Rotary plough feeders are heavy-duty machines used to reclaim stored material. The slot allows the stored material to flow onto a protruding reclaim shelf (Figure 8.12). The design of this shelf prevents the material from flowing over the edge of the reclaim shelf. The ploughing mechanism consists of arms, revolving on a vertical axis, sweeping the stored material from the shelf onto a belt conveyor. The plough feeder can be continuously traversed or positioned at any point along the reclaim shelf for maximum flexibility.

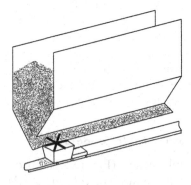

Figure 8.11 Rotary table feeder.   Figure 8.12 Rotary plough feeder.

8.2.10 Drag-scraper feeders

Drag-scraper feeders consist of a succession of plates or bars mounted between two strands of conveyor chain. The plates or bars drag along the bottom of a trough (Figure 8.13). Drag-scraper feeder is a simple and compact arrangement for controlling the feed of fine or small-sized lump materials.

8.3 Chute

Chute is a steeply inclined trough used to feed the material onto the conveyor. In the two types of transfer feeding points, chute becomes essential in between feeder and conveyor in the first type and between two conveyors in the second type. Chute is simple in design and easy to construct and has no moving parts. Chute of rectangular cross section is the most common. The other cross sections are circular and elliptical. Chute does not require power for operation. Maintenance costs are usually low and principally limited to replacement of wear surfaces.

Chute often demands more attention and can be the source of more downtime than the conveyor. Chute, through which the material is fed to the conveyor belt at a transfer feeding point, is to be designed in order to feed the material

1. in the direction of belt travel
2. at the centre of the belt
3. to minimize the impact on the belt
4. at a speed equal to the speed of the belt.

The design of chutes is influenced by the capacity, size and characteristics of material handled, speed and inclination of the belt and whether the belt is fed at single or multiple positions. The following are the six principles of design of chutes:

1. Prevent plugging at impact points
2. Ensure sufficient cross-sectional area
3. Control stream of particles
4. Minimize abrasive wear of chute surface
5. Control generation of dust
6. Minimize particle attrition.

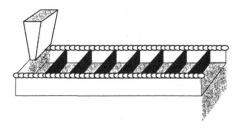

Figure 8.13 Drag-scraper feeder.

8.3.1 Chute angle

The feeding chute must always be inclined to make the material flow at desired forward velocity. If the material is fine with some moisture, the chute must be steep enough that the material will slide rapidly. The chute angle should be low enough to keep lumps not to bounce excessively after they land on the belt. A low chute angle, combined with proper load direction and speed, makes the lumps to impact the belt more gently, thus reducing the risk of damage to the belt and minimizing material degradation and dust generation. But if the chute angle is too low, material flow will slow, leading to a plugged chute.

The chute angle calculation should be based on whether the material is in relatively static condition (when fed from a hopper) or in motion (discharged from another conveyor) and the overall operation. The chute angle is influenced by the lump size, flowability of materials and the atmospheric conditions. The angle of the chute should be 5°–10° more than the angle of repose of the material handled.

8.3.2 Chute dimensions

Chute widths are usually designed to suit the piece of mechanical equipment that follows the chute. The width of a feeding chute should be no longer than two-thirds the width of the receiving belt and not less than 2–3 times the dimension of the largest lump size which is likely to pass through the chute. The cross section or volume of the chute should be at least four times the biggest lump so that the chute does not get blocked. Many times, the chutes are found blocked in spite of all design calculations because of feeding few odd-dimensioned material particles due to negligence.

8.3.3 Chute construction

Feeding chutes can be made of metal or other materials. Metal chutes fabricated from steel plates, typically mild or stainless steel, are the most common. They are available in a wide variety of shapes and sizes, readily fabricated in the shop or field, and are relatively low in cost. For abrasive materials, the chute can be lined with abrasion-resistant, removable plates or other materials, such as ceramic liners. For corrosive materials, corrosion-resistant metal coatings, rubber or synthetic linings can be used.

The chute can be bolted or welded together. The back or bottom plates of the chute should be fitted closely to the belt and provide with rubber edging to prevent leakage of fines. The edging also prevents lumps to block between back plate and the belt. Bolted chutes trade off the advantage of easier assembly/disassembly against the disadvantages that bolted joints are more likely to leak material (due to improper sealing), and the protrusions of the nuts and bolts provide a ready target for material impact, abrasion and corrosion.

8.3.4 Lining of chutes

The lining inside the chutes may be required for a variety of reasons such as one or more of the following:

- To facilitate repair or replacement of the sliding surface
- To improve the performance of an inadequate chute
- To resist abrasive wear
- To resist impact wear, noise, dust generation or product degradation
- To promote flow and avoid build-up, particularly with sticky materials
- To protect against corrosion or chemical reaction by the product handled
- To protect the product from contamination or chemical reaction
- To help to centralize the loading material on the belt
- To preserve the sealing strips from the forces of the material as it loads.

The selection of appropriate lining material for chutes is of fundamental importance to the life and performance of the chute. Rubber, manganese steel, high-chrome steel, chrome–molybdenum steel, plastics, synthetic elastomers and ceramic are some of the lining materials commonly used. Liner initially can be applied to the entire chute surface, or the chute can be used for a short time and liners are then installed at the points where polishing of the metal indicates greater wear.

8.4 Flow aid devices

The obstruction for the material to flow out from the storage vessel, bin, silo, bunker or hopper occurs in the vessel due to

1. the formation of a cohesive arch or bridge
2. the formation of rat-hole
3. the formation of a mechanical arch.

Arching occurs when an obstruction forms above the outlet of the storage vessel and prevents any further material discharge. Rat-holing can occur when the material has sufficient strength to cease flowing into a channel, a stable rat-hole or pipe results. An erratic flow can be the result of alternative arching and rat-holing. Mechanical arch forms when the size of the bulk solid is large in comparison with the outlet of the storage vessel simply by the particles or lumps of material becoming interlocked across the converging section above the outlet.

In such cases, the primitive practices to make the material flow are to beat the sides of the storage vessel with a convenient hammer and stirring or poking the material in the vessel with some kind of rod. Tired of hammering and stirring with rod, engineers began the development of flow aids or discharge aids for bulk materials.

Flow aids or discharge aids are used to return to a more regulated flow pattern and are relatively inexpensive. They are most often used in storage bins and chutes to promote flow of stored material, but they can also be used in continuous-flow process vessels and conveyors.

An ideal discharge flow aid device should have the following:

- It should have a controlled amount of material agitation in the lower regions of the storage vessel, but not provide agitation in the upper regions.
- Maximum efficiency is achieved by transferring energy into the material, not into the vessel walls.

- It would provide the correct amount of vibration so as to provide sufficient particle mobility to assure flow, but not cause compaction.
- It would be quiet in operation.
- The unit would be serviceable externally to the vessel.
- The unit could be retrospectively fitted to any size, shape or configuration of the vessel within a few hours.

The most popular flow aid is the **bin vibrator**: either rotary, pneumatic or electric. Unfortunately, the bin vibrator is very inefficient owing to the way it is installed. This is because the bin vibrator is mounted directly to the wall and, consequently, most of the energy is absorbed by the walls and not by the material in the hopper. When vibration intensity does reach the material, it is often insufficient to be beneficial. As the vibration from a bin vibrator cannot be focused on a particular area of the hopper, vibration can be generated in the upper section of the hopper which causes additional compaction of material in the lower region of the hopper. These units are also rather noisy, as they cause structural vibration of the storage vessel.

The **bin activator** is the most widely used system consisting of a steel dish flexibly supported beneath the vessel opening. A baffle plate, typically in the form of an inverted cone fitted above the opening of the activated section, supports the head load of the stored material and transmits the vibrations directly to it. This bin activator imparts vibration up through the material mass and allows a controlled discharge which is determined by the size of its internal dome/cone and annulus width.

Several varieties of bin activator are commercially available and offer various features which are claimed to improve the discharge characteristics. These bin activators effectively provide a 'live' bottom to the vessel on which they are fitted. Hence, these are popularly called live bottom bin activators. An alternative flow aid device involves fitting a vibrating cage or screen to the inside of the sloping wall of a vessel, with the aim of breaking any arch of product that begins to develop across the converging section.

A simple solution for mechanical bridging or arching is to suspend chains vertically within the vessel so that, if a stable arch develops, an upward pull on the chains should destroy it and restore the flow. Circular bin discharger is the most reliable mechanical dislodger.

Aeration is another solution used to overcome material flow problems. This is where air is introduced into dry normally free-flowing products. **Air cannons** or **air blasters** use compressed air to create an explosive force that is used to dislodge and prevent build-up, blockages and clinging material from the inside walls of the vessel or transfer chute.

8.5 Direction of feeding

Material can be fed onto the conveyor in two possible directions: feeding **in the direction of belt travel** (in-line transfer, Figure 8.14) and feeding **transverse to the direction of belt travel** (non-in-line transfer, Figure 8.15). The first direction, i.e. feeding in the direction of belt travel, really is the best as it facilitates the simplest and easiest design of the chute and the whole transfer feeding point.

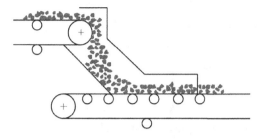

Figure 8.14 In-line transfer point.

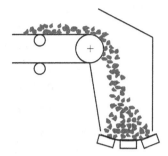

Figure 8.15 Non-in-line transfer point.

Feeding transverse to the direction of belt travel is found very frequently in belt conveyor layout. The horizontal angle that a belt conveyor makes with the succeeding belt may be less or more than 90°. In this case, the design of the chute becomes more complex. Angularity at transfer point needs the turning of flow of material into a new direction. As the angularity increases, chute design becomes more difficult with respect to desirable material velocity and feeding at centre of the belt. There is more wear on the belt cover and increased wear within the chute. Skirtboards may have to be higher and longer to retain the material on the belt without spillage till the material is accelerated to the belt speed and attains proper load shape. It is recommended to avoid angular transfers of more than 90°.

8.6 Centralizing the material

Off-centre feeding of material (material piled deeper predominantly on one side of the conveyor belt) is a problem faced at many transfer points. It is most commonly seen on transfer points which must change the direction of material movement. It is also seen with in-line transfer points (Figure 8.14), particularly when the material is accumulated within the chute, or changes in material characteristics, or changes in material speed at start-up or shutdown, alter the material trajectory and pile material deeper on one side of the conveyor belt. This problem is even more difficult in non-in-line transfer points (Figure 8.15) where the direction of material movement is changed.

If the material does not uniformly spread on the belt, the material accumulates against one of the skirtboards. Then the belt can be displaced transversely (pushed over to one side) of the rollers (Figure 8.16). Such a displacement of the belt leads to belt training problems and may result in spillage of the material over the edge of the belt. Off-centre feeding of the belt makes the belt off-track towards its lightly loaded side.

The chute can help in controlling the load placement by the use of a number of fixtures such as deflectors, baffles, shapers, adjustable flow training gates placed within the transfer chute. These fixtures serve as load centralizers by directing large lumps of material towards the centre of the load where they are less likely to slip off the belt or damage the rubber skirt seal.

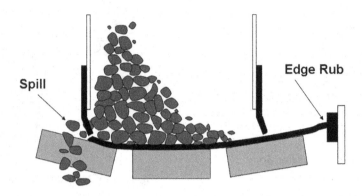

Spill

Edge Rub

Figure 8.16 Effect of off-centre feeding.

If the material to be transported has varying properties with regard to moisture content, particle size, density, etc., the transfer feeding point is fitted with an adjustable deflector (Figure 8.17) within the chute to centre the load.

Installation of adjustable gates allows the direction of the natural flow path into the centre of the receiving belt. These are wood or metal wedges installed on the inside surface of the chute. These movable gates are adjustable from outside the chute to provide easy correction of flow.

8.7 Minimizing the impact

The actual feed point should begin no sooner than the point where the belt is fully troughed and supported by impact rollers or impact cradle or the midpoint of the first set of standard troughing rollers in the impact area.

If feeding is performed while the belt is still in transition, the load is actually dropped onto a slightly larger area with non-parallel sides. This larger area increases

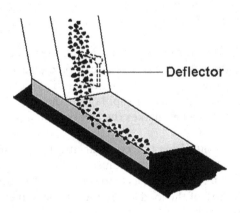

Deflector

Figure 8.17 Deflector inside the chute.

pressure on the side skirts and increases wear of belts and liners. Flooding of material also occurs.

While feeding the material onto the belt, the height of drop of material should be less than 1.5–2 m in order to minimize the impact. However, greater drops are sometimes necessary due to space requirements of crushers, changes in process, equipment, etc.

Very fine materials do not cause much impact when loaded. The rollers under the feeding point are to be closely spaced. Otherwise, belt can be deflected causing leakage under skirtboards. When lumpy materials are fed, impact rollers under the feeding point of the belt should be used to absorb the impact. The wear on the belt cover is great if the material is abrasive or has sharp-edged particles or lumps.

If the material consists of a mixture of fines and lumps, grizzly or screened feeding chute (Figure 8.18) is to be used. Grizzly or screen allows the fines to pass through, forming a bed on the belt. The lumps, unable to pass through, slide down and land on the belt on the top of the bed of fines. Thus, the impact of the lumps is cushioned by the layer of fines. This cushion effect reduces the impact pressure, thereby reducing the damage from the larger lumps.

If the direction of flow of the material feeding onto the conveyor belt is not exactly in the direction of belt travel, some impact is encountered as the material falls on the belt. Hence, the material should be fed onto the conveyor belt in the same direction to that of the belt travel to minimize the impact.

If the speed of the material fed onto the conveyor belt is not equal to the speed of belt travel, the material fed is subject to much turbulence before it accelerates to the speed of the belt. Consequently, the height and length of the skirtboards may have to be increased to retain the material and avoid spillage. If the belt conveyor is inclined (slanting upward), it is difficult to attain the speed of the material equal to the speed of the belt. The difficulty increases with inclination. Hence, it is recommended to make the feeding portion of the belt conveyor horizontal and connect to the succeeding in-clined portion by means of suitable vertical curve. The same recommendation is to be followed in case of declined (slanting downward) conveyors.

A stone box or rock box is used when severely abrasive material is to be fed onto the slow receiving belt to transfer the impact energy to the box rather than the belt. Stone box (Figure 8.19) consists of a chute bottom made as a box. The material

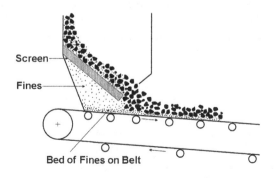

Figure 8.18 Screened feeding chute.

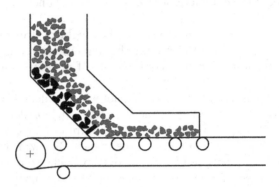

Figure 8.19 Stone box or rock box feeding chute.

flowing through the chute first will retain in this box. The flow of the rest of the material across this box portion is on the retained material. Thus, the wear on the chute bottom is avoided. The other material bounces over the retained material in the box and falls onto the belt. The stone box absorbs the impact from the material flowing over. It can be seen that the stone box also reduces the height of drop of the material.

Multiple rock boxes are also used to reduce the free drop height of material within the chute and allow the material to drop in steps with shorter free drop height at each step. These multiple rock boxes reduce the impact. Usually, the rock boxes are arranged in a way that the material will never have a free drop of more than 5 to 6 feet.

8.8 Material speed equal to belt speed

When a slow-moving material falls on a fast-moving belt, there is a slipping action causing a substantial abrasion to the belt cover. By feeding the material at or near belt speed, this slip is reduced and belt life is improved. The flow of material down a feeding chute onto the moving belt surface can be arranged so that the forward velocity of the material is nearly the same as the belt velocity.

The use of a short, **speed-up belt conveyor** (Figure 8.20) between the feeding chute and the feeding point of a long, high-speed main belt conveyor is one method of feeding the material at a speed nearer to the speed of the main belt conveyor. This is possible as the speed-up belt conveyor runs in the same direction and speed as the main belt conveyor.

The length of the speed-up belt conveyor should be sufficient to bring the speed of the material very close to the speed of the main belt conveyor. The speed-up belt takes the impact and avoids the wear that otherwise would occur to the cover of the main conveyor belt. Thus, it saves the longer and more expensive belt of the main conveyor from damage. The cover of the speed-up belt should be thick enough to accept the wear caused by the acceleration of the material leaving the feeding chute. The speed-up belt also avoids pooling the material on the belt that may be caused due to the difference in the speed of the material and the speed of the belt.

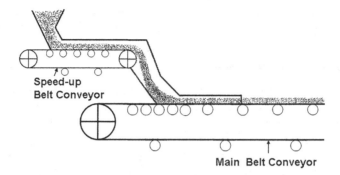

Figure 8.20 Speed-up belt conveyor.

If the impact of the material is high, the speed-up belt conveyor should be flat and should run on flat impact rollers. The flat belt can be made with suitable number of plies and enough thickness of the cover to take the impact. Continuous skirtboards should be provided. If the impact is not severe, a lighter belt may be used with a cover thick enough to take the wear.

8.9 Skirtboards

Skirtboards are also referred to as **skirt plates**, **steel skirting** or sometimes **chutewall**. Skirtboards are strips of metal or wood extending out from the side of the hopper or chute. The two sides extend parallel to one another. Their purpose is to retain the material on the belt after it leaves the feeding chute and until it reaches belt speed and also to keep the material from spilling over the edge of the belt. Skirtboards help to shape and settle the load.

8.9.1 Skirtboard dimensions

Usually, when the feeding is in the direction of the troughed belt travel, the skirtboard length is a function of the difference between the feeding material velocity and the belt velocity. The skirtboards should extend along the belt to two or three times the belt width beyond the feed point where the material has settled and takes the shape it will maintain for the rest of its travel. On level conveyors, skirtboards need be only 3 or 4 ft long, but on rapid or inclined conveyors, they should be longer. On steep inclined conveyors carrying coarse material, skirtboards are sometimes provided for the entire length to prevent spill of any pieces that slide or roll back. Skirtboards are required for the entire length of the conveyor when the load never becomes completely stable. Skirtboards should end above a roller rather than between rollers, as this provides the maximum protection against fluctuations in the line of travel.

The height of the skirtboard must be enough to contain material load under normal operating conditions of the belt. For 20° rollers, the height of the skirtboard is usually one-fifth of the belt width and it is one-third of the belt width for 35° or 45° rollers. Higher skirtboards are required when large lumps are carried by the belt.

Normally, the maximum distance between skirtboards is two-thirds of the width of a troughed belt. Sometimes, especially for free-flowing materials such as grain, the distance between skirtboards is reduced to one-half the width of the troughed belt.

8.9.2 Skirtboard to belt distance

Skirtboards are placed leaving a gap between the lower edge of the skirtboard and the belt. The gap is sealed by a flexible elastomer sealing strip to prevent leakage of fines through the clearance between the lower edge of the skirtboard and the moving belt (Figure 8.21). The sealing strips are bolted or clamped to the exterior of the skirtboard, leaving a gradual widening gap between sealing strip and the belt in the direction of belt travel. This gradual widening gap avoids the material entrapment between the belt and the sealing strip which causes gouging or abrading of the moving belt.

When skirtboard is lifted too much for safety, effective sealing cannot be accomplished and ineffective sealing would result in spillage, material build-up, mistracking of belt, etc. It is easy to maintain a seal between the skirtboard and the belt, if the gap is as small as allowable. Lower edges of skirtboard should be positioned 6 mm above the belt, and this clearance should be uniformly increased in the direction of belt travel to 9 mm. This uniformly increasing gap permits the lumps or foreign objects to move forward without becoming jammed between the lower edge of the skirtboards and the belt.

The rubber used for such an edge sealing should contain no fabric to pick up and retain abrasive particles, thus avoiding the abrasion of the belt cover. Strips of old rubber belting should never be used for edge sealing.

Rubber skirtboard edging should be adjusted frequently so that the edging just touches the belt surface. Forcing the edging hard against the belt cover will not only groove the belt cover, but also require additional power to move the belt. On conveyors with continuous skirtboards, improper pressure of rubber skirtboard edging may overload the belt conveyor driving motor.

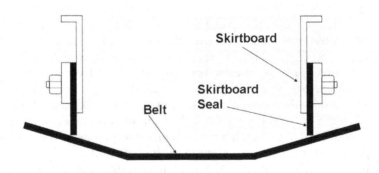

Figure 8.21 Skirtboard seal.

8.10 Wear liners

Wear liner is a material installed on the interior of a skirtboard to be worn away by contact with the moving material. It serves multiple purposes:

1. It provides an easily replaceable wear surface to protect the skirtboard.
2. It prevents application of high side forces on rubber skirting during loading.
3. It helps to centre the material load inside the skirtboards.
4. It provides a free area between the material and the skirting assembly.
5. It can reduce friction, impact, noise and degradation of material.

8.10.1 Wear liner styles

There are three different styles of wear liner commonly used today. They are **deflector**, **straight** and **spaced** wear liners (Figure 8.22).

Deflector wear liner incorporates a bend to provide a free area between the rubber skirting and the wear liner. This free area is useful because fines which get under the wear liner are allowed to travel down the belt to the exit area of the transfer point.

The drawback of the deflector wear liner is that it reduces the effective cross-sectional area of the skirtboard area. This will reduce the volume of material passed through the transfer point and may require adjustments in the chute dimensions to maintain chute capacity. Deflector liner may also reduce the maximum lump size leading to material jam.

Straight wear liner is used to prevent side loading onto the skirting without constricting material flow. It is used on all sizes of the belt. The benefit of straight wear liner is that it provides improved life and sealing effectiveness without closing down the effective load area. Straight wear liner is a good choice to meet the current and future requirements for flow of most materials. Straight wear liner is also best for belts with multiple loading points.

Spaced wear liner is used when the conveyed material requires dust collection. Spaced wear liner is installed slightly away from the skirtboard. Dust collection system draws the air from the area between wear liner and the skirtboard, which is a negative pressure area. Spaced wear liner concept used with a straight or deflector design

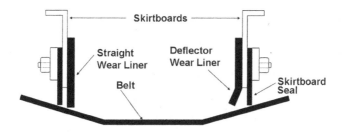

Figure 8.22 Wear liner styles.

resulted in a tremendous improvement over wear liner installed directly on the inside of the skirtboard.

The wear liner materials used are mild steel, carbon steel, stainless steel, abrasion-resistant plate (AR plate), ceramic and ultra-high-molecular-weight (UHMW) polyethylene.

8.10.2 Labyrinth seal

When skirtboard and wear liner are combined with the flexible rubber sealing strip, labyrinth seal is formed. The rubber seal will keep the dust from escaping. It cannot withstand material side pressures. The pocket formed between the deflector wear liner and the sealing strip avoids imposing material side pressure directly on the sealing strip. Fines slipping off and passing through the bottom of the deflector wear liner are contained by the rubber sealing strip. This sealing rubber should be a continuous strip along the sides of the skirtboard. An interlocking or overlapping arrangement is to be made to prevent the leak-out of material through the bottom of adjoining surfaces.

In labyrinth seal (Figure 8.23), wear liner inside the skirtboard and rubber sealing strip with suitable belt support improve the performance of conveyor transfer feeding point and control the escape of material. Rubber sealing strip should be adjusted frequently in such a manner that the edge of the rubber sealing strip just touches the belt surface. If the edge is forced hard against the belt cover, it leads to extra wear of the belt. It causes generation of additional heat due to increased friction and also requires more power to move the belt.

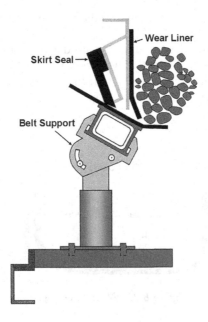

Figure 8.23 Labyrinth seal.

8.11 Methods of discharging from the belt

Material from the belt can be discharged by various ways depending on the characteristics of the material and the purpose of the material discharged. The material is discharged from the belt for different purposes such as temporary storage, long storage
in storage vessels and stockpiles, processing in a plant.

The material can be discharged

1. over an end pulley
2. by discharge chutes
3. by trippers
4. by ploughs.

8.11.1 Discharge over an end pulley

Discharge over an end pulley is the simplest and most common method of discharge
from a conveyor belt. In this method, the material passes over an end pulley and falls
onto a pile (Figure 8.24). When the material is discharged over an end pulley, the
speed of the belt and the diameter of the end pulley determine the trajectory of the
material discharged. Suitable arrangement is to be made not to raise the pile to touch
the end pulley. If pile touches the end pulley, the cover of the belt scrubs against the
top of the pile.

8.11.2 Discharge chutes

Discharge chutes are arranged to collect all the material discharged from the conveyor
belt including the material cleaned by a belt wiper. When the material is discharged
over an end pulley, a suitable discharge chute is added to direct the material to a pile, a
bin or another conveyor. Figure 8.25 shows the discharge chute directing the material
onto another conveyor.

The material discharged over the head pulley of the top conveyor is falling on the
sloping bottom of the chute. This slopping bottom is guiding the material to fall onto
a receiving conveyor belt. It is a good practice to provide a strip of rubber on the upper
edge of this sloping bottom plate, the lip, to eliminate the danger of any hard lump
becoming wedged between the lip and the belt on the pulley, which results in abrading

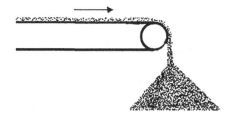

Figure 8.24 Discharge over end pulley.

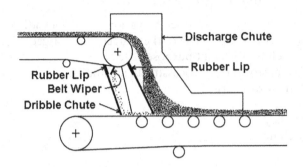

Figure 8.25 Discharge chute.

or gouging of the belt cover. An angled chute called a dribble chute positioned under the head pulley of a top conveyor catches the cleaned material and drops it into the discharge stream. A well-designed chute always allows the free and smooth flow of the discharged material and avoids abrupt changes in material flow direction.

Discharge chutes are also arranged under the hopper of a bin, silo or bunker to discharge the material from the hopper onto the conveyor belt as shown in Figure 8.1.

8.11.3 Trippers

If the material is required to be discharged at any point(s) upstream of the head pulley, the use of trippers is one way. A belt tripper is used to trip the material off the conveyor at specified locations between the terminal pulleys. A belt tripper consists of a structure with two pulley assemblies, one above and forward of the other as shown in Figure 8.26.

As the belt passes over the upper pulley, the material is discharged from the belt into a collection chute. The material is then diverted to one side or both sides of the belt for discharge, or back onto the belt if the desired discharge position is at the end of the conveyor.

Belt trippers are specifically designed to discharge the material from the belt at one or more points or along the length of the conveyor. The chute can be arranged to discharge the material from it to either or both sides of the belt conveyor as required. Figure 8.27 shows the material discharge by tripper to one side, and Figure 8.28 shows the material discharge by tripper to both sides of the conveyor.

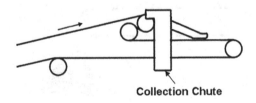

Collection Chute

Figure 8.26 Tripper arranged in a belt conveyor for discharge of material.

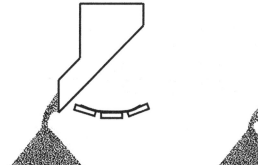

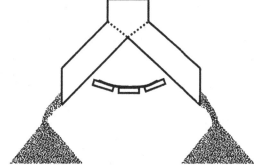

Figure 8.27 Discharge to one side.   *Figure 8.28* Discharge to both sides.

A belt tripper can either be in a fixed position (stationary or fixed tripper) or travel continuously at a constant speed (movable or travelling tripper) for layered stacking. Fixed strippers are less common that discharge materials at the same location. Any number of fixed trippers can be used on a belt conveyor. Figure 8.29 shows belt conveyor discharging the material at three locations: two locations by fixed trippers and one location by an end pulley.

Travelling tripper can be used to discharge material at many points on either side of the belt, or to one side, continuously or intermittently. This tripper moves in forward and reverse directions. Travelling trippers are most commonly used. They are used to load the material from large stockpiles or bunkers and to spread material over a large area along a longitudinal stockpile to increase the storage capacity.

Figure 8.30 shows a belt conveyor discharging material with travelling tripper making a long pile at one side of the belt conveyor. Similar long pile can also be made at the other side of the belt conveyor simultaneously. Travelling tripper can be driven by the conveyor belt itself, or by an electric motor, or by a cable and winch. It can also be locked in any location to act as a fixed tripper.

A fixed tripper can be fitted to carry horizontal reversible cross-belt conveyor. When the belt of the main conveyor to which the tripper is fixed passes over the upper pulley of the fixed tripper, the material is discharged from the belt into a collection chute. This chute can be made to discharge the material from it to one side or both sides depending on the tonnage handled to a reversible cross-belt conveyor. The material thus discharges onto reversible cross-belt get discharges the material to a side in which the

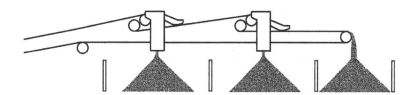

Figure 8.29 Discharge by fixed trippers.

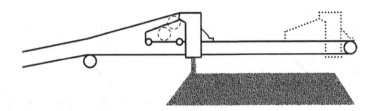

Figure 8.30 Discharge by travelling tripper.

reversible belt is moving. Figure 8.31 shows this kind of arrangement. Individual piles of various materials or grades of the same material can also be stockpiled on either side of the tripper with this arrangement.

If the tripper is travelling tripper in this arrangement, material can be discharged on any side of the belt conveyor in a long pile with the opening of gate corresponding to the travel direction of the cross-belt.

Figure 8.32 shows a fixed tripper outfitted with two inclined transverse belt conveyors on both sides of the conveyor belt to stack the material. These transverse belt conveyors are called stackers. This arrangement helps to build higher piles on either side of the tripper. If the tripper is travelling tripper in this arrangement, it forms long piles on either side of the tripper.

Similar to the travelling tripper with reversible cross-belt, travelling tripper with reversible shuttle belt can be arranged as shown in Figure 8.33. A reversible shuttle

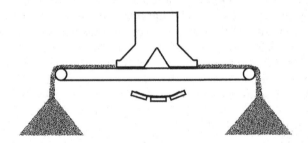

Figure 8.31 Discharge by tripper with reversible cross-belt.

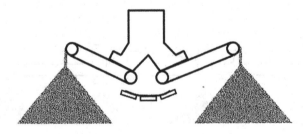

Figure 8.32 Discharge by tripper with two transverse stacker belts.

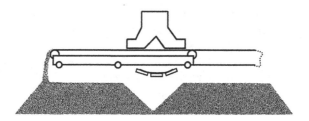

Figure 8.33 Discharge by tripper carrying reversible shuttle belt.

belt conveyor is a belt conveyor that is mounted onto a rail system. The conveyor has the ability to move along the rails in either direction, and the belt can be switched for either direction. The conveyor is half the length of the piles required, giving it plenty of room for several discharge locations. Both functions, belt direction and shuttle movement, can be automated or controlled by an operator. With this arrangement, piles with a large flat top area can be built on either side of the tripper.

8.11.4 Ploughs

Ploughs are rectangular metal blades kept at the desired discharge points on a flat horizontal conveyor belt to direct the material from the belt to either side or simultaneously to both sides of the belt. Material can also be ploughed from flat inclined conveyor belts if the angle of inclination is not too great and material is not too free-flowing. Material can also be discharged by a plough from a troughed belt by temporarily making the belt flattened. Ploughs are either stationary or movable. Movable ploughs cover the discharge points in wider range. Pneumatic or electric actuators are commonly used for operation. Manual lever action can also be arranged.

Ploughs with adjustable rubber bottom edging are set at desired angle to the belt centre line. Different types of ploughs are used for discharging the material to one side, to both sides simultaneously or to both sides simultaneously but with a portion of the material left on the central part of the belt.

Single horizontal swing plough (Figure 8.34) is the single plough swinging horizontally across the belt on a vertical pivot and is used to discharge the material to one side. Horizontal movement of the plough facilitates the removal of all the material or part of the material from the belt. **Single lift plough** (Figure 8.35) is the single plough lifting vertically on a horizontal pivot and is used to discharge the material to one side. In a

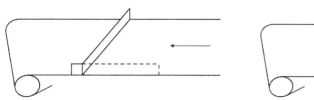

Figure 8.34 Single horizontal swing plough.

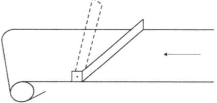

Figure 8.35 Single lift plough.

single lift plough, plough is lowered to discharge the material from the belt or raised to avoid discharge. In both types, ploughs can be manually operated or powered by a pneumatic or hydraulic cylinder to remove all or part of the material from the belt.

Ploughs discharging the material simultaneously to both sides of a flat belt conveyor are called V-ploughs. **Horizontal V-plough, vertical V-plough, proportioning V-plough** and **travelling V-plough** are the four types of V-ploughs used for removing the material from the flat belt. In any V-plough, two blades form V shape. In a horizontal V-plough, individual blades swing horizontally on vertical pivots as shown in Figure 8.36 and join at the centre. This horizontal V-plough removes part of the material to one side and another part to the other side. In a vertical V-plough, fixed V shape formed from two blades is angularly raised or lowered as shown in Figure 8.37 to remove all the material. Both the V-ploughs may be manual- or power-operated.

In a proportioning V-plough, two blades form an open V shape as shown in Figure 8.38 allowing some of the material to remain on the central portion of the belt while discharging outer portions of the material to both sides of the belt. Amount of opening at the apex of the V can be made adjustable depending upon the requirement. A travelling V-plough shown in Figure 8.39 discharges the material to both sides in any number of specific positions. It is usually operated by cable and winch in either direction. This plough is used to discharge the material from the troughed belts. A flat belt support plate is used to lift the belt from the troughing rollers to make the belt flat in order to discharge the material from the belt by travelling V-plough.

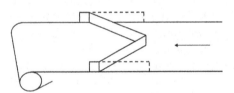

Figure 8.36 Horizontal V-plough.

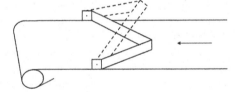

Figure 8.37 Vertical V-plough.

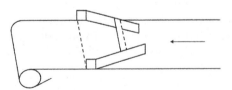

Figure 8.38 Proportioning V-plough.

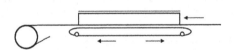

Figure 8.39 Travelling V-plough.

Chapter 9

Spillage and dust control

Spillage is the material lost that has fallen from the sides of the conveyor belt, typically at the feed zone, but can occur at any point along the conveyor. Spillage is a general term for all fugitive material. Dust is any solid particle smaller than 500 μm in diameter that can remain airborne. Obviously, dust is also a part of spillage. Therefore, control of the spillage can control the dust originated from the spillage. But dust comes out from the material while it is handled needs to be controlled. Hence, control of spillage and dust has to be dealt separately.

9.1 Spillage

Spillage is a dreaded problem for all conveyor users. Spillage is highly deleterious to the belt and the rollers, which are not only critical but also expensive accounting for nearly 50% of the total cost of the belt conveyor system. To regulate the cost of material handling, it is to be aimed to get the maximum life and trouble-free performance of these components.

Erratic running of the belt is one of the major causes for material spillage. A belt that runs true minimizes the spillage. The quality of the material conveyed affects the true running of the belt to a major extent. Two categories of materials, viz. fine dry and dusty material and wet and sticky material, appear to be the most problematic in relation to the generation of dust, mess and spillage.

Material can spill or escape from the handling path at the following four crucial points in a conveyor system:

1. The feed point
2. The upper strand (conveying run)
3. The discharge point
4. The lower strand (return run)

At the feeding chute, some material will spill or spurt off the belt. This problem gets accentuated especially at the feeding points of sloping conveyors. The problem with fine dry material is that it readily becomes airborne and also escapes at the transfer points. Feeding the belt before it is troughed fully, flattened impact rollers and improper skirt rubber sealing are some of the causes for spillage.

Feed chute should load the material on to the belt at a speed equal to the speed of the belt and in the direction the belt travel. Otherwise, the spillage is liable to occur. Tight seal at the feeding point can be created with snug fitting skirting starting at the chute and running the length of the belt. Belt should be supported properly at the feeding point. If not, the belt will sag when the material is loaded on the belt. The sag will break the seal and allow spillage to occur.

Feeding zone spillage is best controlled by a three-part programme consisting of the following:

- Proper belt support to minimize belt sag
- Wear liners inside the chute to protect the sealing system
- Multiple-layer edge seals to contain any escaping fines

On the upper strand of the belt, any material that is deposited unevenly or is heaped up is liable to spill over the sides. Such spillage will also occur from troughed belts that do not run true and tilt sideways. If the belt ascends too steeply, individual large pieces of materials roll back and get lodged, down the slope or jump over the side of the conveyor. Wrong concave or convex curves in an ascending conveyor cause spillage, till the belt rests on the rollers. The belt should run over the rollers as smoothly as possible. Otherwise, the belt jerks up and down and throws the material out as spillage and also causes dust.

At the discharge head pulley, spillage may occur due to choking or blockage by too large pieces of material. This may be due to poor design, or bad workmanship on chute, or careless patchwork or repairs done at site. Gaps or holes in the chute are the major source of spillage. The dust enclosure or guard, where the belt leaves the feeding zone, should be in such a way that they allow minimum outside air.

Most material escapes from the return run of the belt. Material that is not thrown off the belt at the head pulley, or fails to drop off by gravity, adheres to the belt in varying degrees depending on the particle size and percentage of fines, plus the degree of moisture in the material. Fine moist material clings to the carrying surface of the belt and is not discharged with the material at the discharge points. This material is carried back sticking to the return run of the belt, and some of it detaches from the surface of the belt as it proceeds on to the snub/bend pulley surfaces and then on to the return rollers. Later on, the carryback material falls off when dried out or dislodged by vibration of the return rollers. Such carryback material causes environmental hazards, belt carcass rupture in the long run, run-off or snaking of belt, edge damage, difficulty to run the belt centrally, etc. Most of the material collects near the beginning of the return run, more particularly near the head pulley, the take-up pulley or the first return rollers.

Only a practical engineer who has suffered will be aware of the magnitude of the spillage problem. There are installations where expenditure on clearing and cleaning the spillage is well over 3%–5% of the unit cost of production. One of the criteria to control this spillage problem is to have a minimum number of conveyors in a system, especially when they are connected in sequence. A close and careful study of the conveyor in operation will provide not only vital clues to the spillage problem but also solutions.

9.1.1 Causes of spillage

The spillage can occur due to the following:

- Faulty transfer points
- Loading the material transverse to the direction of the belt travel
- Loading the material at a speed less than the speed of the belt travel
- Irregular feeding that makes the belt empty and overloaded portions alternately, which leads to spillage of material over the edges of the belt along the overloaded portions
- Absence of skirtboards or insufficient dimension of skirtboards
- Ineffective sealing between skirtboards and the belt
- Conveyor with steep inclination
- Belt snaking or lining out of belt
- Damaged chutes or holes in chute plates, due to wear, missing bolts and gap between plates at the junction points
- Ineffective or insufficient cleaning mechanism
- Wind
- Moisture content in the material
- Nearness to sea, resulting in lifting off the belt, or tendency to push the belt to one direction

Generally, spillage starts from the feeding section. Hence, it would be worthwhile to spend more time and effort in observing the problem and taking corrective action, rather than accepting this as an inevitable evil. This problem would need careful attention and discussion, from the planning stage and taking action at the time of the following:

- Drawing the specification
- Selection of equipment
- Erection of equipment
- Trial run of the equipment
- Takeover of the equipment, after correcting the deficiencies

The extra care exercised will be paid back multifold in later years.

9.1.2 Effects of spillage

Fugitive material, spillage and/or dust, escaping from conveyors, are an everyday occurrence in many plants. Plant's personnel have come to accept spillage and dust from leaky transfer points and other sources within the plant as the routine and unalterable course of events. But the problem is that fugitive material represents an ever-increasing drain on a conveyor's efficiency, productivity and profitability.

Material lost from the conveyor system costs the plant in a number of ways. The following are some of those areas.

9.1.2.1 Reduced material-handling efficiency

The most expensive material is the material spilled from the belt. When the material becomes fugitive, it has to be gathered up and reintroduced into the system. This adds additional handling, which increases the overhead, and therefore, the cost of production. If the fugitive material is lost and cannot be salvaged, it even more dramatically reduces the efficiency.

9.1.2.2 Reduced plant safety

Many conveyor-related accidents are a direct or indirect result of fugitive material created by the conveyor. Recent statistics indicate that roughly one-half of the accidents that occur around belt conveyors in mines are attributable to clean-up and repairs necessitated by spillage and build-up. If fugitive material could be eliminated, the frequency at which personnel are exposed to these hazards would be significantly reduced.

9.1.2.3 Diminished quality

Fugitive material can contaminate the plant, the process and the finished product in several ways, thereby diminishing the quality of the product. Material reclaimed from the floor may have suffered degradation or contamination. Material can be deposited on sensitive equipment and influence sensor readings.

9.1.2.4 Increased conveyor maintenance costs

As spilled material escapes from a conveyor system, it leads to a number of problems in the conveyor itself. These problems increase maintenance expenses for service and the replacement of components.

1. The first and most visible expense is the cost of clean-up. This includes the cost for the man-hours for personnel to clean and return the material to the belt, and also equipment hours if used. As the personnel have their attention diverted to clean-up, other maintenance activities may be delayed.
2. Fugitive material accumulates on various conveyor components and other nearby equipment. Rollers can jam when clogged or buried under fugitive material. Once jammed, the constant wear of the belt will eventually cut through the roller shell. This leaves a razor-shaped edge that can badly damage the soft underside of the belt. Jammed rollers and pulleys also increase the friction drag that can dramatically increase the power consumption of the conveyor.
3. Even if the rollers are not buried in fugitive material, the presence of fugitive material in the environment makes the roller to stop rolling as the material migrate through the seal to the roller bearing. Then, the belt will wear a hole in the roller. Even moderately worn rollers have flat spots that cause uneven rotation, excessive belt wear and unnecessary friction that wastes conveyor power.
4. Rollers facing fugitive material are replaced prematurely, in advance of their expected wear life. This brings added costs in the form of additional equipment expenses, the labour costs for installation, and the cost of downtime or loss of production to perform the required installations.
5. The material build-up on the face of pulley and rollers can force the belt to run off-centre. If it runs completely off the pulley, it can rub and even fold over into the conveyor structure. If this is not noticed right away, the damage can destroy great lengths of valuable belt.

6. Material accumulation can lead to significant belt training problems, which leads to damage to the belt and other equipment (as well as the risk of injury to the personnel). This belt wander also creates interruptions in production as the belt must be stopped, repaired and retrained prior to resuming operations.

All the above problems become self-perpetuating: spillage leads to build-ups on rollers, which leads to belt wander, leading to more spillage.

9.1.2.5 Lowered employee morale

Physical environment at workplace influences a worker's feelings. Employees have better morale when the workplace is clean and neat. Workers with higher morale are more likely to be at work on time and to perform better in their assignments. People tend to feel proud if their place of work is a showplace, and they will work to keep it that way.

9.1.3 Steps to control spillage

The necessary steps to be taken in ensuring that a plant is reasonably dust-, mess- and spillage-free are as follows:

1. Make the right choice of handling system
2. Ensure good design of equipment based on correct technical specifications
3. Ensure proper design of belt support and sealing in the load zone area
4. Ensure central loading on conveyor belt
5. Ensure feeding the material on to the belt at a speed equal to the speed of the belt and in the direction of belt travel
6. Install belt training devices to align the belt and keep it running true
7. Fit appropriate dust entrainment and belt cleaning equipment
8. Ensure ease of clean-up, e.g. by adequate access
9. Implement regular systematic maintenance schemes and close follow-up
10. Educate personnel in general about the following:
    • good housekeeping
    • wastage of good material
    • importance of pollution prevention, etc.

9.2 Dust

Dust consists of tiny solid particles carried by air currents. These particles are formed by disintegration force such as impact. A wide range of particle size is produced during a dust generating process. Particles that are too large to remain airborne settle, while others remain in the air indefinitely.

Dust is broadly defined as any particle smaller than 500 µm in diameter that can remain airborne. Particles of 10 µm or smaller in diameter are classified as respirable dusts. Respirable dust refers to those dust particles that are small enough to penetrate the nose and upper respiratory system and deep into the lungs. Particles that penetrate deep into the respiratory system are generally beyond the body's natural clearance mechanisms of cilia and mucous and are more likely to be retained. When suspended in air, respirable particulate matter affects humans as it is inhaled, forcing the heart and lungs to work harder to provide oxygen to the body. This can lead to a decreased breathing ability and damage to the heart. Particles of more than 10 µm, called inhalable dust, if inhaled by human, are trapped in the nose, throat and upper respiratory tract.

Although unavoidable in bulk material-handling plants, the escape of dust particles into the workplace atmosphere is undesirable. Excessive dust emissions can cause both health and industrial problems. Few of them are as follows:

- Health hazards
  - Occupational respiratory diseases
  - Irritation to eyes, ears, nose and throat
  - Irritation to skin
- Impaired visibility
- Unpleasant odours
- Risk of dust explosions and fire
- Damage to equipment
- Material loss
- Increased maintenance requirements
- Decreased productivity

In addition to the above, risks to employees include safety hazards associated with manual clean-up and working in dirty and dusty environments. Outside the plant, the risks extend to the surrounding community and environment that are exposed to the pollution and health concerns arising from the dust.

Hence, more stringent environmental regulations are forcing many material-handling plants to substantially reduce dust emissions. Because of this, users and designers are looking at the options available to them to meet the codes of environmental regulations. Many facilities have realized that by improving the transfer of material between conveyor belts, the amount of dust can be reduced significantly, making the entire facility and operation healthier for employees and the public at large.

Properly designed, maintained and operated dust control systems can reduce dust emissions and, thus, workers' exposure to harmful dusts. Dust control systems can also reduce equipment wear, maintenance and downtime; increase visibility; and boost employee morale and productivity.

The saying "prevention is better than cure" can certainly be applied to the control of dust. Although total prevention of dust is an impossible task, properly designed belt conveyor components can play an important role in reducing dust generation, emission and dispersion.

Belt conveyors emit dust from the following four points:

- The tail end, where material is received
- The conveyor skirting
- The head end, where material is discharged
- The return rollers, due to carryback of fine dust on the return belt

9.2.1 Dust prevention measures

The following are the few dust prevention measures to be taken at various parts/points of belt conveyor:

- Bin or hopper is to be completely enclosed to minimize the dust emission
- Care is to be taken to select the right feeder that produces less agitation of the material to minimize the dust generation, and it should be enclosed as much as possible to minimize the dust emission
- At feeding points, the material is to be fed onto the centre of the belt, at a speed equal to the speed of the belt and in the direction of belt travel to minimize the dust generation
- In a transfer chute, abrupt changes in direction are to be avoided to reduce the possibility of material build-up, material jamming and dust generation
- Stone box is to be arranged to place a layer of fines ahead of the lumps on the belt, which helps to prevent heavy impact of material on the belt, to reduce the belt wear and dust generation
- Impact rollers at feeding point are to be adequately spaced to absorb the force of impact, to prevent the deflection of the belt between the rollers and to prevent dust leakage under the rubber seal
- Skirting design is to be improved with inclined rubber strips as dust seal to allow the material to rest on the moving belt at all times even when the belt is momentarily deflected between rollers
- A dribble chute is to be provided to redirect the material removed by the belt scraper into the process stream to reduce the carryback of fine materials on the return belt
- Dust leakage falling on the non carrying side of the belt is to be cleaned by V-plough cleaner to prevent build-up of material and dust on the tail pulley
- Dust curtains are to be used to contain dust within a conveyor enclosure. They are segmented rubber or plastic curtains (baffles) suspended inside an enclosed duct to slow down airflow and allow airborne dust to settle back into the material stream on a conveyor belt. They should be installed at the head, tail and exit ends to provide easy access during maintenance

Lowering chutes are used for directing the fragile, degradable and/or dusty material to storage from the discharge point of the belt conveyor. They are effective dust control equipment and also prevent degradation of the conveying material. **Spiral lowering chute**, **bin lowering chute**, **rock ladder** or **stone box** and **telescopic chute** are four common forms of lowering chutes.

9.2.1.1 Spiral lowering chute

Figure 9.1 shows the usual form of a spiral lowering chute used for gently lowering fragile and/or dusty bulk materials to prevent dust generation.

9.2.1.2 Bin lowering chute

Bin lowering chute is used to direct the material to storage bin without generating large amounts of dust. This chute consists of a channel that runs from the discharging

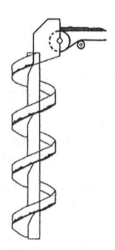

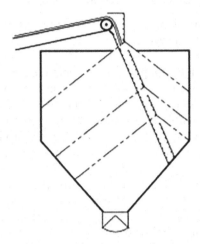

Figure 9.1 Spiral lowering chute. (Courtesy: CEMA.)

Figure 9.2 Bin lowering chute. (Courtesy: CEMA.)

equipment down into the storage bin. The channel is secured to the sloping side of the bin near the bin bottom. Material slides down the chute quietly, with minimum dusting. When the material meets the bin side or the surface of the material in the bin, it leaves the chute from the sides and spreads out conically (Figure 9.2). When properly arranged, this bin lowering chute will eliminate dusting and fill the bin to its practical capacity.

9.2.1.3 Rock ladder or stone box

The rock ladder is a structural steel (or wood) tower having a series of baffles arranged in such a manner that bulk material discharged from a belt conveyor never has a free drop of more than 5 ft. This lower drop height significantly reduces the dust (Figure 9.3). If the material is heavy, abrasive and lumpy, the baffles can be arranged in the form of stone box.

9.2.1.4 Telescopic chute

The telescopic chute is used to minimize the height of the material fall into the stockpile, thereby minimizing dusting. Telescoping sections usually are cable connected in such a manner that a winch will successively lift the sections of the chute. The lower end of the chute is always kept just clear of the top of the stockpile to reduce dust (Figure 9.4). Coal is often delivered to storage piles through telescopic chutes.

9.2.2 Dust control

After all the necessary preventive measures have been adopted, the dust still remaining in the workplace or particularly at transfer point can be controlled by one or more of the following techniques:

Figure 9.3 Rock ladder or stone box. (Courtesy: CEMA.)

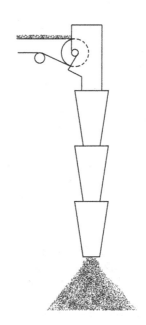

Figure 9.4 Telescopic chute. (Courtesy: CEMA.)

- Dust containment systems
- Dust collection systems and
- Wet dust suppression systems

9.2.2.1 Dust containment systems

Dust containment systems mechanically control and confine the dust on the conveyor or transfer point. Airflow through the transfer point can be reduced by controlling the amount of air entering the transfer point, by building the enclosure large enough to slow or minimize air speed and by utilizing additional control measures to slow air movement. As air velocity is reduced, airborne particles that are too heavy to be supported by the reduced air speed drop down on the material from the air stream.

The sources of the dust include belt carryback, conveyor side spillage, tail area spillage and exit area dust creation. Each source must be identified and corrected as part of a total dust containment system.

9.2.2.2 Dust collection systems

There are passive and active dust collection systems available. In a passive system, the air carrying the dust is allowed from the transfer point to a filter where the dust particles are collected. In an active system, the air carrying the dust is pulled under vacuum through a filter to collect the dust particles. Mechanical dust collection systems are installed to pull dust-laden air away from the conveyor transfer point, separate the dust and then exhaust clean air. A typical dust collection system consists of four major components:

- Exhaust hoods or pickups, to capture dust at the transfer point
- Duct work to transport the captured air and dust mixture to a collector
- A collector, filter or a separation device to remove die dust from the air
- A fan and motor to provide the necessary exhaust volume of energy

9.2.2.3 Wet dust suppression systems

Wet dust suppression systems use liquids (usually water) with or without chemicals to wet the material so that the weight of each dust particle is increased and they are less likely to become airborne. As the fines are wetted, they agglomerate with increasing cohesive force forming larger and heavier groups of particles making more difficult to carry by the air. The most common method of wetting the material is through a series of properly sized and located nozzles at a discharge point of the conveyor belt where the material expands and moves in air.

The following wet dust suppression systems are most commonly used to control the dust at the transfer point:

- Plain water spray or water cannons
- Water surfactant spray
- Water/air surfactant/spray
- Fog spray or fog cannons
- Foam spray

Chapter 10

Belt training

Normally, the belt will not wander on a well-engineered, aligned and maintained conveyor structure. But, in reality, belt wanders or runs off-track. This wandering of the belt causes material spillage, component failure and costly damage to the conveyor belt itself. Few of the reasons for wandering or snaking of the belt, to the left or to the right, are indicated in Section 5.2.5.

In most cases, belt wander can be prevented by careful and correct alignment of conveyor structure and components. Conveyor stringers must be centred and in line. All pulleys and rollers must be levelled, installed on the conveyor centre line and perpendicular to the belt centre line.

In case of off-centre feeding, the centre of gravity of the load will seek the lowest point of the troughing rollers, leading the belt to wander on it lightly loaded edge (Figure 8.16). Accumulation of fugitive material other than at the centre of the rollers and pulleys makes the belt run that side where material is accumulated; thus, the belt is wandered. It is corrected by proper feeding chute arrangements.

Accurate squaring of the belt ends prior to splicing is essential for belt mistracking not to take place, and it helps distribute stress evenly throughout the splice. Belt wandering due to not squaring the belt ends can be corrected by re-splicing the belt with accurate squaring of the belt ends. A concave curve of the belt is known as **bow**. A bow can be created during defective manufacture of the belt.

When non-uniform wear of lagging of pulleys makes mistracking of the belt, replacement of the pulley with uniform lagging would be the solution. Heavy equipment operators are to be cautioned and trained to operate the equipment with greater care to avoid structural damage. Ground subsidence can easily be found during routine checks. The structure should be levelled with proper foundation.

When strong winds on one side of the conveyor cause enough lateral force to move the belt off-line, either the conveyor is to be enclosed or windbreak is to be provided on the wind side. Similarly, if the belt moves off-line because of rain, ice, snow or heat, some form of conveyor cover would be the solution.

When the belt is wandered or runs off-track, it is to be trained to run 'true'. Belt training, or tracking, is defined as the procedure required to make the conveyor belt run 'true' when empty and also when fully loaded.

The basic rule of training is that "the belt moves towards that end of a roller which it contacts first". When one end of the roller is higher than the other, the belt will climb the roller; as it lays over the top of the roller, it contacts the high end first. Training the belt is a process of adjusting rollers, pulleys and loading conditions in a manner that will correct any tendencies of the belt to run other than true.

The belt is trained by adjusting the rollers near the location where it is wandering. The adjustment should be spread over some length of the conveyor preceding the region of trouble. The movement of one roller generally has its greatest training effect in the area within 5–8 m downstream of that roller. Only one roller is to be shifted at a time, as shifting additional rollers may cause over-correction. Making slight adjustment is always good rather than extreme ones.

After making an adjustment, run the belt to evaluate its effect, but wait for two to three belt revolutions before making further adjustments, as the effects of roller movement are not always immediately apparent.

The best procedure to use in starting the training sequence is to start with the return run and work from the head (discharge) pulley towards the tail (feed) pulley. This will ensure easy centring of the belt as it comes around the tail pulley for centralized feeding. Then follow with the top run in the direction of belt travel. Start with the belt empty. After training is completed, run the belt with full load to check the results.

If a check shows the belt has been over-corrected, it should be restored by moving back the same roller rather than shifting additional rollers. It should be noted that before the belt is put on a full production schedule, all training adjustments should be finalized.

To reiterate, the general sequence for training is as follows:

1. Return run, from head to tail
2. Carrying run, from tail to head
3. First empty, then loaded, then with the belt running

In starting to train the belt, priority should be given to those points where, if the belt should run off, damage may occur if the conveyor is not stopped. When it has been assured that no damage can be caused to the belt, accurate training can begin.

10.1 Knocking and tilting the rollers

Knocking and **tilting** the rollers are two methods for training the belt. Shifting the roller axis with respect to the path of the belt is commonly called knocking the rollers. This is effective when the entire belt runs to one side along some portions of the conveyor.

The belt can be centred by knocking ahead the end of the rollers in the direction of belt travel. If the belt is mistraining to the right, the left side of the frame is fixed and the right side is slightly advanced. This is to be done for one or two rollers before the point of detraining. It is better to adjust several rollers minutely rather than one roller excessively.

Knocking is to be done only if the belt makes good contact with all three troughing rollers. Obviously, such roller shifting is effective for only one direction of belt travel. If the belt reverses, a shifted roller which is corrective in one direction becomes misdirective in the other, worsening the problem. Hence, reversing belts should have all rollers squared up and left that way. Any correction required should be provided with self-aligning rollers designed for reversing belts.

Shifting the position of rollers may have benefits in belt training. However, there are drawbacks as well. Rollers knocked in all directions in an effort to train the belt create additional friction, resulting in belt cover wear and increased power consumption.

By tilting the roller slightly forward in the direction of belt travel, additional pressure is placed on standard roller or self-aligning roller where strong training influence is required. This is accomplished by inserting a shim underneath the rear leg of the roller frame. This shim, which can be as unsophisticated as a steel washer, sets up centring forces that will help to keep a belt running straight.

However, too much tilt can reduce the contact between belt and roller or cause excessive wear on the pulley side of the belt. This method has an advantage over knocking rollers in that it will correct erratic belts which track to either side. It has the disadvantage of encouraging pulley cover wear through increased friction on the troughing rollers. Consequently, it should be used sparingly, especially on higher-angle troughing rollers. The effect of tilting the roller is counterproductive on a reversing belt.

10.2 Mechanical training aids

The following are the mechanical training aids used to train the belt:

1. Self-aligning carrying training roller
2. Self-aligning return training roller
3. V return roller
4. Multiple-axis training roller

The training effect of all these rollers is already described in Section 5.2.5. In addition to the above rollers, Crowned-face pulleys and misalignment switches are also used for training the belt.

Crowned-face pulleys provide a centralizing effect and make the belt to run true. Crowns are most effective where there is a long unsupported span, four times the belt width or greater, approaching the pulley. As this is not often possible on the carrying side of the conveyor, head pulley crowning is relatively ineffective and is probably not worth the stress it produces in the belt.

Belt misalignment switches are installed near the outer limit of safe belt travel and act to shut off the power to the conveyor in the event the belt moves past this safe limit. They usually consist of a roller or lever installed at both edges of the belt. When the belt moves too far in either direction, it physically contacts the lever, pushing it over past its limit and breaking the power circuit. The belt stops, and operator must realign the belt.

Another method to centre the belt as it approaches the tail pulley is to slightly advance and raise alternate ends of the return rolls nearest to the tail pulley. The theory here is that the competing forces of the deliberately induced mistracking in opposite directions works to centre the belt.

The training aids, viz. self-aligning and V return rollers, crowned-face pulleys, belt alignment switches, can never be totally effective to prevent belt wandering. Strictly speaking, conveyor operations should not depend on training aids to overcome gross misalignments of the conveyor structure. Continuous activity will overwork the training aids and reduce their longevity. If any of the training aids are working continuously, it means that there is a basic training problem to be identified and corrected as soon as possible.

Chapter 11

Operation, maintenance and safety

For the belt conveyor system, operation, maintenance and safety are interlinked. Unless all possible safety measures are taken and all the parts of the belt conveyor are properly maintained, the operation of the belt conveyor cannot be successful. Hence, it is very essential to develop an operating procedure and to offer a programme of instruction for all personnel involved in operation, maintenance and safety of a belt conveyor system at the time of installation and during trial runs. These programs are to be repeated, and procedures are to be reviewed at frequent intervals for updating the knowledge of trained personnel and motivating the new personnel.

Proper access to each and every part of the belt conveyor system by operating and maintenance personnel is very critical to its productivity. Insufficient access to equipment results in loss of production and dirty systems, due to the difficulty of cleaning and making required repairs. Providing proper access for easy to see, easy to reach, easy to inspect and easy to replace is a fundamental consideration for effective operation and maintenance of the belt conveyor system. Walkways and work platforms beside the conveyor provide a firm path adjacent to the conveyor and around head and tail pulleys, with easy access to all points where observation, lubrication, clean-up and other maintenance activities are required.

11.1 Operation

A well-designed and installed belt conveyor can be operated by one or two operators. An operator can monitor the performance of the belt conveyor with the help of electrical controls, built-in safety sensors and devices, closed-circuit TV and other signal systems from a remote area. One or two mechanics with a supervisor, trained for the purpose, can easily patrol the system at regular intervals to detect the condition of a component that needs attention.

Two of the most important factors in the successful operation of a belt conveyor are proper belt tension by the take-up assembly and training the belt to run safely within the rollers and returns to avoid costly downtime, damage to belt edge and spilling of material.

Various operation-based sensors, such as belt slip sensors, belt tracking or misalignment sensors, plugged chute sensors, metal detectors, will monitor the belt during its operation to detect when the belt will be out of normal operating conditions so that the operator can minimize or avoid damage to the belt because of deviation from the normal operating conditions. Measurement of conveyor belt cover wear rate using ultrasonic gauges, eddy-current-based sensors and laser-based scans helps the operator to predict the expected remaining life of the belt. X-ray-based and magnetic scanning of belt and belt splice reveals the damage to the fabric or steel card reinforcement and splice deformation. Material spillage sensor is useful to detect longitudinal rip, a long tear or cut, in a conveyor belt allowing quick action to be taken to protect the belt from further damage. Variation of the belt speed detected by the speed monitors and variation of the belt tensions detected by counterweight limit sensors are helpful in smooth operation of the belt conveyor.

11.2 Maintenance

Maintenance can be defined as the set of activities on all machineries of the conveyor system, to maintain the same at prime condition in economic ways, for smooth running of the belt conveyor for the achievement of organizational goals with respect to targets and to meet the standards of environment and safety. Prime condition refers to that the machinery shall be in good condition for efficient and effective utilization of the same for the functions for which it is designed and installed for.

Maintenance practices will help to control fugitive materials, improve safety and reduce unscheduled conveyor outages. Routine maintenance inspections can extend the life of the belt and its components and improve the belt performance. Efficient and effective maintenance lowers costs, not just for maintenance department, but for the total operation. Only competent and trained personnel, equipped with proper test equipment and tools, can be allowed to perform conveyor maintenance.

Housekeeping and lubrication of bearings are the two important routine maintenance activities in a conveyor system. Good housekeeping to clear the fugitive material and clean the machinery and surroundings forms the low-cost maintenance. In addition, both exterior and interior of every machine should be kept free from dust, oil and moisture. Dust can have a harmful effect if it is allowed to settle on the windings, to enter the bearing or to collect in the ventilating passages. Motors should be blown out periodically depending on the atmospheric conditions at site. Due to the large number of bearings present in a conveyor system, lubrication of bearings as per the manufacturer's recommendations will enhance life expectancy of rolling components.

Maintenance is an integral part of production activities of an organization. To put in a nutshell, the **fivefold objectives** of systematic maintenance of belt conveyors could be briefly summarized as detailed below:

1. To keep the belt conveyor system running at maximum productive efficiency, consistent with quality and reliability
2. To keep the transportation costs to the minimum, by aiming for **zero downtime**
3. To reduce the actual maintenance workload to the minimum (i.e. to achieve maximum availability at economic cost)
4. To improve the performance of the conveyor system while in operation
5. To ensure the **safety** of the conveyor equipment and the personnel through improved working conditions (for the operators) and maintenance efficiency.

In short, optimum and efficient utilization of the available resources, viz. **5 Ms –men, materials, machinery, methods, money** – and last but not the least **time**, is the basic foundation of a sound system.

The following are the three main types of maintenance practices adopted in a belt conveyor system:

1. Breakdown maintenance
2. Preventive maintenance
3. Predictive maintenance

In **breakdown maintenance**, the equipment is run till it fails and then repaired. This is also referred to as **repair maintenance**. It is not a scientific approach at all, since the only attention the conveyor equipment receives is at the time of breakdown or failure. Also, such equipment breakdowns, which may recur and increase at an alarming frequency, would cause tremendous production loss, reducing the safety and reliability of the equipment. Hence, as far as possible, this practice is discouraged and avoided.

In **preventive maintenance**, the emphasis is on the adoption of measures to anticipate breakdowns/failures and to prevent them before they occur, rather than allowing them to occur and then take remedial action. Each and every work to be carried out on conveyor is examined, analysed and pre-planned in detail so that there may not be any delay in executing the same. Preventive maintenance includes lubrication programmes.

In **predictive maintenance**, the equipment is no longer disassembled, as a routine of inspection, but instead, the exact condition or behaviour of this productive equipment is diagnosed by means of **on-stream non-destructive testing**. The use of instruments such as **sensors, amplifiers, analysers, read-out units** enables us to detect incipient malfunction, if any, well in advance, allowing for planned shutdowns and stable production control. Even lubricant contaminations can be effectively studied through **spectroscope examinations,** and probable malfunctions can be detected. Detection of crack on conveyor structures and flaws in pulley shafts and machined parts of conveyors is now possible by **ultrasonic methods**. Monitoring instruments are available even to check the wall thickness of pulleys, rollers, etc., so that timely action can be taken for their renewal, well before the failure/breakdown occurs. The use of sensitive **flaw detectors** insure against breakdowns and makes predictive maintenance possible at the most appropriate time. Hence, this condition monitoring technique is used extensively on high-speed and large-capacity belt conveyor system with great advantage and is strongly recommended.

A combination of preventative and predictive maintenance programmes will help to ensure maximum availability and component life. A well-maintained conveyor system should be able to consistently operate with 90% mechanical availability.

In Table 12.1 of Troubleshooting chapter, belt conveyor operational troubles, probable causes and corrective measures are listed. It can be a good reference in training the personnel and serve as a maintenance instruction.

It should be remembered that all moving machinery should be stopped during the periods of lubrication, maintenance or adjustment, and be sure that the equipment cannot be restarted during such procedures.

11.3 Safety

Conveyors can be one of the most hazardous mine or plant equipment installations if safety regulations are not strictly followed or if the conveyors are not properly maintained. Good common sense is the key to working on any equipment and must be used while observing or servicing equipment. Conveyor safety begins with a design that avoids foreseeable hazards. The careful consideration of safety aspects during manufacture and installation, and the implementation of proper operating and maintenance policies, all will help to promote a safe working environment on site.

However, many accidents in connection with belt conveyors are caused more by human carelessness, negligence or ignorance, than by the shortcomings of design. Operating and maintenance personnel should therefore receive adequate training on safety matters and should only then be duly authorized to enter and work in hazardous area. No work should be carried out on a conveyor unless the equipment is stationary and electrically locked off.

Good housekeeping is a prerequisite of safety. The conveyor area, access ways and walkways should be kept free of material spillage, surplus equipment, belt trimmings and all unnecessary obstructions. Good lighting, weather protection and non-slip walking surfaces all pay handsome dividends to safety. It is good practice to institute regular visual inspections by responsible engineers to verify the well-being of plant safety.

Guarding is very much essential especially for rollers, pulleys, drive components and shafts to protect the personnel from potential injury from these components. In addition to guarding these components, labelling the guarding also helps the personnel to identify the specific hazards associated with the guard.

Electrical accessories are electrically operated devices which are not required for normal starting and stopping of belt conveyor. Pre-start alarms, pull cord switches, blocked chute devices, bin level switches, misalignment switches, belt weighing scales and zero-speed switches are few of the electrical accessories used in most of the belt conveyor systems to alert the personal for possible care.

All conveyors must be equipped with pre-start alarms with an audible and visual system that provides a pre-start warning along the entire length of the conveyor to alert the personnel working at or near the belt conveyor. These include horns, sirens, flashing lights or strobes, or a combination of two or more warning signals. Where belt conveyors are able to be accessed from both sides, the pull cord switches are located on both sides of the conveyor with latching attachment. An operator activates the switch by pulling the cord until the switch trips, interrupting power to the conveyor and usually raising a visual indicator flag. The switch remains tripped until reset manually at the switch location. The belt does not restart till the switch is reset. A blocked chute device provides belt protection at the discharge end of the conveyor into a transfer chute. Actuation of the plugged chute switch with time delay normally results in the tripping of the conveyor drive. Blocked flow can result in damage to the moving conveyor. A blocked chute can also cause severe damage to the belt being fed, particularly in the case of a single large lump stuck in the feeding boot and slitting the belt. Plugged chute switches are used in many configurations depending on the application.

When conveyors discharge into bins or hoppers, bin level sensors provide protection to the belt in that they shut down the conveyor if the predetermined level is exceeded. Misalignment switches are often located on the unsupported section of the belt in a horizontal take-up system in order to minimize the damage that misalignment can do in this area. Switches consist of roller switches, limit switches, whisker switches, proximity switches or photoelectric switches. Conveyor belts are sometimes protected from overload by belt weighing scales that measure the belt loading at a given point. A zero-speed switch is connected to some rotating member of a conveyor to cut off power to the conveyor if the conveyor speed falls below a predetermined minimum.

Spill protection is a guard provided to catch and hold lumps of bulk materials, build-ups of frozen material or failed rollers that may fall from the conveyor onto walkways, roadways or buildings making a hazard to personnel or equipment below. Spill protection is often arranged at designated cross under locations and work areas located near or under conveyors. A shallow trough called drip pan is fitted underneath a belt conveyor to prevent carryback or water from falling to the surface below.

Chapter 12

Troubleshooting

Any belt conveyor installation can be subject to a wide variety of troubles that may become costly. Those troubles can result in replacement and plant downtime unless the problem is quickly diagnosed and corrected.

Troubles usually arise because of the following three types of faults:

1. **Faults with the belt or its splices**

    A. Belt

    i. The belt is bowed, cambered or cupped

    A concave curve of the belt is known as **bow**. A bow can be created during manufacture, improper storing, splicing or tensioning.

    A convex curve of the belt is known as **camber**.

    The action of the edges of the belt curving upward on the carrying side and downward on the return run is called **cupping**. It is also referred to as belt **curl**. Depending on severity, it may not be able to be guided by the rollers or even be loaded properly.

    ii. There are defects or damage in the carcass (plies or cards) of the belt

    iii. The belt edge or cover is damaged

    iv. There is belt degradation from exposure to the elements or to chemicals

    B. Manufacturing or application

    i. The belt is poorly matched to the structure or application

    ii. The belt has a 'bow' or 'camber' from its manufacturing process

    iii. The belt was not stored properly

    C. Splices

    i. There was poor installation of a vulcanized or mechanical splice resulting in a splice that is not square to the belt

    ii. A belt was formed from several pieces joined all the wrong ends resulting in a camber or crooked section

    iii. Different types, thicknesses or widths of belt have been spliced together

    iv. The belt has splices that are damaged or coming apart

2. **Faults with the conveyor structure, components or the environment**

    A. Structure

    i. The structure was not accurately aligned during its construction

    ii. The structure has settled on one side through ground subsidence

iii. The structure has been damaged from plugged chutes, fires or collisions with mobile equipment

B. Components

   i. Rolling components (rollers and pulleys) are not aligned in all the three axes
   ii. The gravity take-up is misaligned
   iii. Rolls of a roller have seized or been removed
   iv. Material build-up or wear has altered the profile of rollers or pulleys

C. Environment

   i. The conveyor is subjected to high winds
   ii. Rain, frost or ice and snow build-up altered the friction on one side of the belt
   iii. The sun shines on one side of the conveyor

3. **Faults with material loading**

   i. The load is not centred on the receiving belt
   ii. The load is segregated with larger lumps on one side of the belt
   iii. There is intermittent loading on a belt that is tracked for a constant load

In Table 12.1, the majority of the troubles encountered while the belt conveyor is in operation are given. Probable causes are indicated, and corrective measures are suggested. This is a guide to the reader to understand the interrelationship between various components of the belt conveyor to set forth probable causes and corrective measures whenever a trouble arises.

Table 12.1 Belt conveyor troubles, probable causes and corrective measures

Trouble	Probable causes	Corrective measures
Belt runs to one side at a particular point along the conveyor	Rollers or pulleys out-of-square with the centre line of the belt	Readjust rollers in the affected area; install limit switches for greater safety
	Rollers improperly placed	Relocate rollers or insert additional rollers spaced to support the belt
	Jam/frozen/sticking rollers	Free rollers; improve maintenance and lubrication; replace if necessary
	Loose roller	Secure roller in proper position
	Conveyor frame or structure crooked	Stretch string along edge to determine the extent and make correction
	Conveyor structure not level, belt climbs to high side	Align and level the conveyor structure
	Material build-up (on pulleys and rollers)	Remove accumulation; install cleaning devices, scrapers and inverted V decking; use skirtboards properly; improve maintenance and housekeeping

(Continued)

Trouble	Probable causes	Corrective measures
Belt runs to one side for long distance or the entire length of conveyor	Belt running off-centre around the tail pulley and through the loading area	Install training rollers on the return run prior to tail pulley
	Improper belt loading	Adjust chute and loading conditions so that the load is centred and it is in the direction of belt travel; control flow with chutes, feeders and deflectors to load the material at the speed of the belt
	Rollers or pulleys out-of-square with the centre line of the belt	Readjust rollers in the affected area; install limit switches for greater safety
	Material build-up (on pulleys and rollers)	Remove accumulation; install cleaning devices, scrapers and inverted V decking; use skirtboards properly; improve maintenance and housekeeping
	Conveyor frame or structure crooked	Stretch string along edge to determine the extent and make correction
	Roller stands not centred on belt	Readjust rollers in the affected area
Belt runs to one side throughout the entire length at specific rollers	Rollers or pulleys out-of-square with the centre line of the belt	Readjust rollers in the affected area; install limit switches for greater safety
	Rollers improperly placed	Relocate rollers or insert additional rollers spaced to support the belt
	Material build-up (on pulleys and rollers)	Remove accumulation; install cleaning devices, scrapers and inverted V decking; use skirtboards properly; improve maintenance and housekeeping
	Jam/frozen/sticking rollers	Free rollers; improve maintenance and lubrication; replace if necessary
	Conveyor structure not level, belt climbs to high side	Align and level the conveyor structure
Belt runs to one side at some distance along the conveyor	Improper belt loading	Adjust chute and loading conditions so that the load is centred and it is in the direction of belt travel; control flow with chutes, feeders and deflectors to load the material at the speed of the belt
Particular section of belt runs to one side at all points on the conveyor	Belt not joined squarely or used wrong fasteners	Cut ends square and re-splice; use correct fasteners; set up regular inspection schedule

Trouble	Probable causes	Corrective measures
	Belt edge worn or broken	Repair belt edge; remove worn or out-of-square section and splice in new section
	Bowed belt	Avoid telescoping belt rolls or storing them in damp locations; for new belt, bowing should disappear during break-in; in rare instances, belt must be straightened or replaced
	Crooked textile belt	If the textile belt is new, it might straighten out after a couple of days under full load
Belt runs true when empty, but mistracks when loaded	Improper belt loading	Adjust chute and loading conditions so that the load is centred and it is in the direction of belt travel; control flow with chutes, feeders, and deflectors to load the material at the speed of the belt
	Belt not making good contact with all rollers	Adjust height of rollers for even contact of the belt
Belt runs off at the head pulley	Rollers or pulleys out-of-square with the centre line of the belt	Readjust rollers in the affected area; install limit switches for greater safety
	Pulley lagging worn	Replace worn pulley lagging; use grooved lagging for wet conditions; tighten loose and protruding bolts
	Material spillage and build-up (on pulleys and rollers)	Improve loading and transfer conditions; remove accumulation; install cleaning devices, scrapers and inverted V decking; use skirtboards properly; improve maintenance and housekeeping
	Rollers improperly placed	Relocate rollers or insert additional rollers spaced to support the belt
	Head pulley or troughing rollers leading to head pulley out of alignment	Survey and align pulley and adjacent troughing rollers
	Roller stands not centred on belt	Readjust rollers in the affected area
	Pulleys too small	Use larger-diameter pulleys
Belt runs off at the tail pulley	Counterweight too light	Recalculate the weight required and adjust counterweight or screw take-up tension accordingly
	Rollers or pulleys out-of-square with the centre line of the belt	Readjust rollers in the affected area; install limit switches for greater safety

(Continued)

Trouble	Probable causes	Corrective measures
	Jam/frozen/sticking rollers	Free rollers; improve maintenance and lubrication; replace if necessary
	Improper belt loading	Adjust chute and loading conditions so that the load is centred and it is in the direction of belt travel; control flow with chutes, feeders and deflectors to load the material at the speed of the belt
	Material spillage and build-up (on pulleys and rollers)	Improve loading and transfer conditions; remove accumulation; install cleaning devices, scrapers and inverted V decking; use skirtboards properly; improve maintenance and housekeeping
	Belt running off-centre around the tail pulley and through the loading area	Install training rollers on the return run prior to tail pulley
	Dirty, sticking or misaligned return rolls	Clean, lubricate, align return rollers; install cleaning devices; use self-cleaning return rolls
	Return rollers out of line	Adjust at right angle to the frame
	Pulley lagging worn	Replace worn pulley lagging; use grooved lagging for wet conditions; tighten loose and protruding bolts
Belt runs off on certain days	Environmental conditions alter tracking	Install wind screens and/or conveyor covers, and self-aligning rollers
Entire belt runs off at all points on conveyor	Improper belt loading	Adjust chute and loading conditions so that the load is centred and it is in the direction of belt travel; control flow with chutes, feeders and deflectors to load the material at the speed of the belt
	Rollers or pulleys out-of-square with the centre line of the belt	Readjust rollers in the affected area; install limit switches for greater safety
	Material build-up (on pulleys and rollers)	Remove accumulation; install cleaning devices, scrapers and inverted V decking; use skirtboards properly; improve maintenance and housekeeping
	Belt strained on one side	Allow time for new belt to break in. If belt does not break in properly or is not new, remove strained section and splice in a new piece.

Trouble	Probable causes	Corrective measures
	Rollers improperly placed	Relocate rollers or insert additional rollers spaced to support the belt
	Self-aligning (guide) roller sets are stuck (jammed)	Make the self-aligning roller sets movable and ensure that they remain movable or replace them.
	Skirts improperly placed or not maintained	Install skirtboards so that they do not rub against the belt.
Particular section of belt runs off at all points on conveyor	Belt not joined squarely or used wrong fasteners	Cut ends square and re-splice; use correct fasteners; set up regular inspection schedule
	Belt edge worn or broken	Repair belt edge; remove worn or out-of-square section and splice in new section
	Bowed belt	Avoid telescoping belt rolls or storing them in damp locations; for new belt, bowing should disappear during break-in; in rare instances, belt must be straightened or replaced
	Pulley lagging worn	Replace worn pulley lagging; use grooved lagging for wet conditions; tighten loose and protruding bolts
Belt runs off-centre around the tail pulley and in the loading area	Misaligned pulley or belt camber	Align the pulley
	Dirt build-up on pulley	Clean pulley; install training rollers
Belt has erratic action, following no certain pattern	Belt too stiff to train	If new belt, allow break-in time; use more troughable belt; install self-aligning rollers; increase tension/conform to crowns
Belt slip on starting	Insufficient traction between belt and pulley	Lag the drive pulley; in wet conditions, use grooved lagging; increase the belt wrap; install correct belt-cleaning devices
	Counterweight too light	Recalculate the weight required and adjust counterweight or screw; take-up tension accordingly
	Insufficient counterweight travel	Maintain recommended minimum travel distances
	Pulley lagging worn	Replace worn pulley lagging; use grooved lagging for wet conditions; tighten loose and protruding bolts

(Continued)

Trouble	Probable causes	Corrective measures
	Pulleys too small	Use larger-diameter pulleys
	Drive under-belted	Recalculate maximum belt tensions and select correct belt
	Improper initial position of counterweight carriage leading to getting stuck in its guide and cause apparent excessive belt stretch	Identify and remove the cause responsible for the counterweight carriage getting stuck in its guide
	Material between belt and pulley	Use skirtboards properly; remove accumulation; install cleaners and ploughs; improve maintenance
Belt slip	Insufficient traction between belt and pulley	Lag the drive pulley; in wet conditions, use grooved lagging; increase the belt wrap; install correct belt-cleaning devices
	Counterweight too light or blocked	Add counterweight or increase take-up tension or eliminate blocking
	Pulley lagging worn	Replace worn pulley lagging; use grooved lagging for wet conditions; tighten loose and protruding bolts
	Pulleys too small	Use larger-diameter pulleys
	Material spillage and build-up (on pulleys and rollers)	Improve loading and transfer conditions; remove accumulation; install cleaning devices, scrapers and inverted V decking; use skirtboards properly; improve maintenance and housekeeping
	Jam/frozen/sticking rollers	Free rollers; improve maintenance and lubrication; replace if necessary
	Belt overloaded	Operate belt feed system at design capacity or less
Excessive belt stretch	High belt tension	Reduce belt tension; install belt with high permissible tension; use lagged pulley; consider soft or slower start-up
	Belt strength too low	Replace by stronger belt
	Improper belt installation causing apparent excessive belt stretch	Pull belt through the counterweight with tension equal to least empty running tension; run the belt with mechanical fasteners initially
	Drive under-belted	Recalculate maximum belt tensions and select correct belt

Trouble	Probable causes	Corrective measures
	Counterweight too heavy	Recalculate the weight required and adjust counterweight accordingly; reduce take-up tension to point of slipping and then tighten slightly
	Improper initial position of counterweight carriage leading to getting stuck in its guide and cause apparent excessive belt stretch	Identify and remove the cause responsible for the counterweight carriage getting stuck in its guide
	Insufficient counterweight travel	Provide sufficient counterweight travel
	Material build-up (on pulleys and rollers)	Remove accumulation; install cleaning devices, scrapers and inverted V decking; use skirtboards properly; improve maintenance and housekeeping
	Wrong differential speed on dual pulleys	Make necessary adjustment
	Damage by abrasives, acid, chemicals, heat, mildew, oil, etc.	Use belt designed for specific conditions; make spot repairs; seal metal fasteners or replace with vulcanized step splice; do not over-lubricate rollers; enclose belt line for protection against rain, snow or sun; use mildew inhibitor on the belt
	High-elongation carcass	Replace by low-elongation belt, for instance with polyester warp
	Splice starting to open	Dangerous; check splices (X-ray)
	Pulley lagging worn	Replace worn pulley lagging; use grooved lagging for wet conditions; tighten loose and protruding bolts
Excessive belt top cover wear, including rips, gouges, ruptures and tears	Extreme material impact	Use correctly designed chutes or baffles; install impact rollers; make vulcanized splices; adjust skirting; load fines first
	Relative loading velocity too high or too low	Adjust chutes or correct belt speed; consider the use of impact rollers
	Improper belt loading	Adjust chute and loading conditions so that the load is centred and it is in the direction of belt travel; control flow with chutes, feeders and deflectors to load the material at the speed of the belt

(Continued)

Trouble	Probable causes	Corrective measures
	Material build-up (on pulleys and rollers)	Remove accumulation; install cleaning devices, scrapers and inverted V decking; use skirtboards properly; improve maintenance and housekeeping
	Dirty, sticking or misaligned return rolls	Clean, lubricate and align return rollers; install cleaning devices; use self-cleaning return rolls
	Breaker strip missing or inadequate	Install belt with proper breaker strip
	Cover quality too low	Replace with belt of heavier cover gauge or higher-quality rubber
	Excessive sag between rollers causing load to move and shift on belt as it passes over rollers	Increase tension if too low; reduce roller spacing; install belt support rails to maintain sag-free belt line
	Scraper pressed too hard on belt	Position the scraper correctly; do not use old belt as scrapper
	Belt rubbing over heap of spilled/leaked material below the conveyor	Prevent spillage/leakage of material; improve housekeeping
	Abrasive skirtboards	Use rubber skirt material; avoid the use of old belting
	Skirts improperly placed or not maintained	Install skirtboards so that they do not rub against the belt.
	Damage by abrasives, acid, chemicals, heat, mildew, oil, etc.	Use belt designed for specific conditions; make spot repairs; seal metal fasteners or replace with vulcanized step splice; do not over-lubricate rollers; enclose belt line for protection against rain, snow or sun; use mildew inhibitor on the belt
	Belt folding back on self	Remove obstruction causing belt edge to fold back
	Pulley lagging worn	Replace worn pulley lagging; use grooved lagging for wet conditions; tighten loose and protruding bolts
Excessive wear on belt bottom cover	Material build-up (on pulleys and rollers)	Remove accumulation; install cleaning devices, scrapers and inverted V decking; use skirtboards properly; improve maintenance and housekeeping
	Material grinding between pulley and belt	Install tail protection ploughs and scrapers in front of tail pulley
	Insufficient traction between belt and drive pulley leading to belt slippage	Lag the drive pulley; in wet conditions, use grooved lagging; increase the belt wrap; install correct belt-cleaning devices

Trouble	Probable causes	Corrective measures
	Transition length too short	Check and extend, if possible
	Breaker strip missing or inadequate	Install belt with proper breaker strip
	Excessive troughing roller tilt	Correct to not more than 2° from upright
	Jam/frozen/sticking rollers	Free rollers; improve maintenance and lubrication; replace if necessary
	Pulley lagging worn	Replace worn pulley lagging; use grooved lagging for wet conditions; tighten loose and protruding bolts
Excessive belt edge wear, broken edges	Improper belt loading	Adjust chute and loading conditions so that the load is centred and it is in the direction of belt travel; control flow with chutes, feeders and deflectors to load the material at the speed of the belt
	Belt strained on one side	Allow time for new belt to break in. If belt does not break in properly or is not new, remove strained section and splice in a new piece
	Bowed belt	Avoid telescoping belt rolls or storing them in damp locations; for new belt, bowing should disappear during break-in; in rare instances, belt must be straightened or replaced
	Material build-up (on pulleys and rollers)	Remove accumulation; install cleaning devices, scrapers and inverted V decking; use skirtboards properly; improve maintenance and housekeeping
	Belt hitting or rubbing conveyor structure	Install training rollers on carrying and return runs
	Damage by abrasives, acid, chemicals, heat, mildew, oil, etc.	Use belt designed for specific conditions; make spot repairs; seal metal fasteners or replace with vulcanized step splice; do not over-lubricate rollers; enclose belt line for protection against rain, snow or sun; use mildew inhibitor on the belt
Grooving, cracking, gouging or stripping of belt top cover	Skirts improperly placed; skirt seals are stiff and pressed against belt; material entrapped by skirting	Install skirtboards so that they do not rub against the belt; use skirtboard rubber for seal; adjust skirts with gap increasing in the direction of belt travel

(Continued)

Trouble	Probable causes	Corrective measures
	Belt spanks up and down on impact, allowing material to become wedged into skirting	Install impact cradle, rollers or troughing rails under loading zone to maintain stable belt line
	Material hanging up in or under chute	Improve loading to reduce spillage; install baffles; widen chute
	Extreme material impact	Use correctly designed chutes or baffles; install impact rollers; make vulcanized splices; adjust skirting; load fines first
	Material build-up (on pulleys and rollers)	Remove accumulation; install cleaning devices, scrapers and inverted V decking; use skirtboards properly; improve maintenance and housekeeping
	Sharp edges of material or tramp iron coming in contact with cover	Use impact rollers and magnetic removal equipment
	Jam/frozen/sticking rollers	Free rollers; improve maintenance and lubrication; replace if necessary
Grooving, cracking, gouging or stripping of belt bottom cover	Jam/frozen/sticking rollers	Free rollers and improve maintenance and lubrication; replace if necessary
	Material spillage and build-up (on pulleys and rollers)	Improve loading and transfer conditions; remove accumulation; install cleaning devices, scrapers and inverted V decking; use skirtboards properly; improve maintenance and housekeeping
	Slippage on the drive pulley	Increase tension through screw take-up or add counterweight; lag the drive pulley; increase arc of contact
	Pulley lagging worn	Replace worn pulley lagging; use grooved lagging for wet conditions; tighten loose and protruding bolts
	Excessive sag between rollers causing load to move and shift on belt as it passes over rollers	Increase tension if too low; reduce roller spacing; install belt support rails to maintain sag-free belt line
Longitudinal grooving or cracking of belt top cover	Skirts improperly placed or not maintained	Install skirtboards so that they do not rub against the belt
	Jam/frozen/sticking rollers	Free rollers and improve maintenance and lubrication; replace if necessary

Trouble	Probable causes	Corrective measures
	Material build-up (on pulleys and rollers)	Remove accumulation; install cleaning devices, scrapers and inverted V decking; use skirtboards properly; improve maintenance and housekeeping
	Extreme material impact	Use correctly designed chutes or baffles; install impact rollers; make vulcanized splices; adjust skirting; load fines first
Longitudinal grooving or cracking of bottom cover	Jam/frozen/sticking rollers	Free rollers; improve maintenance and lubrication; replace if necessary
	Material build-up (on pulleys and rollers)	Remove accumulation; install cleaning devices, scrapers and inverted V decking; use skirtboards properly; improve maintenance and housekeeping
	Pulley lagging worn	Replace worn pulley lagging; use grooved lagging for wet conditions; tighten loose and protruding bolts
	Slippage on the drive pulley	Increase tension through screw take-up or add counterweight; lag the drive pulley; increase arc of contact
	Excessive sag between rollers causing load to move and shift on belt as it passes over rollers	Increase tension if too low; reduce roller spacing; install belt support rails to maintain sag-free belt line
Longitudinal rip or cut partially or entirely through belt	Damage by external body or jammed conveyor part	Usually cold or hot repair required; mechanical fasteners may be used; replacement of belt is unavoidable
	Jam of material or tramp iron	Improve loading situation; install magnetic separator; install rip detection system
	Defective rollers fell and belt dragged over steel edge	Replace and fix roller on the support
	Rupture of pulley rim	Replace pulley
	Fastener failure	Improve splice or vulcanize
Belt cover swells in spots or streaks Belt covers become checked or brittle	Damage by abrasives, acid, chemicals, heat, mildew, oil, etc.	Use belt designed for specific conditions; make spot repairs; seal metal fasteners or replace with vulcanized step splice; do not over-lubricate rollers; enclose belt line for protection against rain, snow or sun; use mildew inhibitor on the belt

(Continued)

Trouble	Probable causes	Corrective measures
Belt fabric decay, carcass cracks, ruptures, gouges, soft spots	Extreme material impact	Use correctly designed chutes or baffles; install impact rollers; make vulcanized splices; adjust skirting; load fines first
	Material build-up (on pulleys and rollers)	Remove accumulation; install cleaning devices, scrapers and inverted V decking; use skirtboards properly; improve maintenance and housekeeping
	Breaker strip missing or inadequate	Install belt with proper breaker strip
	Drive under-belted	Recalculate maximum belt tensions and select correct belt
	Radius of convex vertical curve too small	Increase curve radius to prevent excessive edge tension; decrease roller spacing in curve; lower any elevated rollers in curve
	Damage by abrasives, acid, chemicals, heat, mildew, oil, etc.	Use belt designed for specific conditions; make spot repairs; seal metal fasteners or replace with vulcanized step splice; do not over-lubricate rollers; enclose belt line for protection against rain, snow or sun; use mildew inhibitor on the belt
	Top rubber cover is of inadequate thickness	Reduce maximum lump size or fit heavier belt or increase belt cover thickness; modify feed chute design
Crosswise cracks of carcass or textile plies, while covers are intact	Low pulley diameters causes extreme flexing of belt carcass	Increase pulley diameter; install more flexible belt
	Material trapped between belt and tail pulley	Install tail protection plough to sweep material from return run of the belt; install decking or guards to prevent material falling onto return run
Belt breaks at or behind fasteners; fasteners pull out or tear loose	Belt not joined squarely or used wrong fasteners	Cut ends square and re-splice; use correct fasteners; set up regular inspection schedule
	High belt tension	Reduce belt tension; install belt with high permissible tension; use lagged pulley; consider soft or slower start-up
	Pulleys too small	Use larger-diameter pulleys
	Pulley lagging worn	Replace worn pulley lagging; use grooved lagging for wet conditions; tighten loose and protruding bolts

Trouble	Probable causes	Corrective measures
	Material build-up (on pulleys and rollers)	Remove accumulation; install cleaning devices, scrapers and inverted V decking; use skirtboards properly; improve maintenance and housekeeping
	Belt carcass too light	Select stronger carcass
Belt hardens or cracks	Damage by abrasives, acid, chemicals, heat, mildew, oil, etc.	Use belt designed for specific conditions; make spot repairs; seal metal fasteners or replace with vulcanized step splice; do not over-lubricate rollers; enclose belt line for protection against rain, snow or sun; use mildew inhibitor on the belt
	Pulleys too small	Use larger-diameter pulleys
	Pulley lagging worn	Replace worn pulley lagging; use grooved lagging for wet conditions; tighten loose and protruding bolts
	Improper storage or handling	Follow proper storage and handling instructions
Belt covers harden or crack	Cover compound too hard	Cover compound contains too many sulphur bridges. No repair possible
	Exposure to heat; Temperature of conveyed material is too high	Use high-temperature-resistant belt type
	Heat or chemical damage	Use the belt designed for the specific condition
	Improper storage or handling	Follow proper storage and handling instructions
Holes or breaks in belt	Impact of material on belt, or foreign body clamped between belt and pulley	Reduce impact; use cushion rollers; use belt with impact protection system; repair the spots
Carcass or cover fatigue at roller junction	Improper transition between troughed belt and terminal pulleys	Increase transition length; use transition rollers; elevate terminal pulley
	Excessive sag between rollers causing load to move and shift on belt as it passes over rollers	Increase tension if too low; reduce roller spacing; install belt support rails to maintain sag-free belt line
	Radius of convex vertical curve too small	Increase curve radius to prevent excessive edge tension; decrease roller spacing in curve; lower any elevated rollers in curve
	Excessive troughing roller tilt	Correct to not more than 2° from upright
	Insufficient belt stiffness	Replace belt with adequate stiffness

(Continued)

Trouble	Probable causes	Corrective measures
Short breaks in carcass parallel to belt edge, star breaks in carcass	Extreme material impact	Use correctly designed chutes or baffles; install impact rollers; make vulcanized splices; adjust skirting; load fines first
	Material between belt and pulley	Use skirtboards properly; remove accumulation; install cleaners and ploughs; improve maintenance
Belt top cover softens	Material contains oil or grease	Use oil-and-grease-resistant cover
	Lubricant has come in contact with belt surface	Do not over-lubricate items; check grease seals; clean belt
Severe pulley cover wear	Slippage on the drive pulley	Increase tension through screw take-up or add counterweight; lag the drive pulley; increase arc of contact
	Material spillage and build-up (on pulleys and rollers)	Improve loading and transfer conditions; remove accumulation; install cleaning devices, scrapers and inverted V decking; use skirtboards properly; improve maintenance and housekeeping
	Material trapped between belt and tail pulley	Install tail protection plough to sweep material from return run of the belt; install decking or guards to prevent material falling onto return run
	Bolt heads protruding above lagging	Tighten bolts; replace lagging; use vulcanized lagging
Fasteners pull out of belt	Wrong type of fasteners or fasteners not tight	Use correct fasteners; use vulcanized splices if feasible
	High belt tension	Reduce belt tension; install belt with high permissible tension; use lagged pulley; consider soft or slower start-up
	Mildew	Use mildew inhibitor on belt
	Improper starting	Use more acceleration steps in starting
Belt edge damage	Bad belt training causing belt to mistrack and run into steelwork	Critical with steel cord belts; adjust pulleys and rollers for straight tracking; repair severe damage by hot vulcanization; realign belt empty and loaded
	Radius of convex vertical curve too small	Increase curve radius to prevent excessive edge tension; decrease roller spacing in curve; lower any elevated rollers in curve
Transverse breaks at belt edge	Belt edges folding up on structure	Install limit switches; provide more clearance

Trouble	Probable causes	Corrective measures
	Improper transition between troughed belt and terminal pulleys	Increase transition length; use transition rollers; elevate terminal pulley
	Radius of convex vertical curve too small	Increase curve radius to prevent excessive edge tension; decrease roller spacing in curve; lower any elevated rollers in curve
Belt edges lift off rollers	Bottom cover too thin or carcass thickness too low	Install belt with thicker carcass or bottom cover
	High belt tension	Reduce belt tension; install belt with high permissible tension; use lagged pulley; consider soft or slower start-up
Belt damage in centre	Belt pulled into nip point between rolls	Change roller frames; increase gauge length
	Transitions incorrect	Introduce additional rollers and/or increase the length of transition zones
Ply or cover separation	High belt tension	Reduce belt tension; install belt with high permissible tension; use lagged pulley; consider soft or slower start-up
	Pulleys too small	Use larger-diameter pulleys
	Belt edge worn or broken	Repair belt edge; remove worn or out-of-square section and splice in new section
	Damage by abrasives, acid, chemicals, heat, mildew, oil, etc.	Use belt designed for specific conditions; make spot repairs; seal metal fasteners or replace with vulcanized step splice; do not over-lubricate rollers; enclose belt line for protection against rain, snow or sun; use mildew inhibitor on the belt
	Belt speed too fast	Reduce belt speed
	Too many reverse bends	Use more flexible belt
	Insufficient transverse stiffness	Replace with the proper belt
	Mildew	Use mildew inhibitor on belt
	Lack of adhesion	Check whether adhesion between cover and carcass of the belt is sufficient
	Cover is damaged, and fine material penetrates into belt	Repair cuts in covers which prevent penetration of fines into the carcass
	Heat or chemical damage	Use belt designed for specific conditions
Vulcanized splice separation	Belt improperly spliced	Re-splice using proper method
	Pulleys too small	Use larger-diameter pulleys

(Continued)

Trouble	Probable causes	Corrective measures
	High belt tension	Reduce belt tension; install belt with high permissible tension; use lagged pulley; consider soft or slower start-up
	Material trapped between belt and tail pulley	Install tail protection plough to sweep material from return run of the belt; install decking or guards to prevent material falling onto return run
	Improper transition between troughed belt and terminal pulleys	Increase transition length; use transition rollers; elevate terminal pulley
	Drive under-belted	Recalculate maximum belt tensions and select correct belt
	Damage by abrasives, acid, chemicals, heat, mildew, oil, etc.	Use belt designed for specific conditions; make spot repairs; seal metal fasteners or replace with vulcanized step splice; do not over-lubricate rollers; enclose belt line for protection against rain, snow or sun; use mildew inhibitor on the belt
Build-up on bend pulleys and return rollers	Insufficient number of belt cleaners	Install additional belt cleaners or maintain existing cleaners more frequently
	Bulk material properties have changed	If there is a permanent change in bulk materials, redesign chutes, belt cleaners and re-evaluate conveyor speed, tension and belt type
Spillage of fines and small particles in loading area	Skirts improperly placed or not maintained	Install skirtboards so that they do not rub against the belt
	Wear liners missing, worn or improperly installed	Replace wear liners so that the bottom edge is lined up and gradually relieving in the direction of belt travel.
	Improper belt loading	Adjust chute and loading conditions so that the load is centred and it is in the direction of belt travel; control flow with chutes, feeders and deflectors to load the material at the speed of the belt
	Excessive belt sag	Recalculate take-up tension; install belt support systems; reduce roller spacing
Spillage of larger particles and lumps along conveyor	Rollers or pulleys out-of-square with the centre line of the belt	Readjust rollers in the affected area; install limit switches for greater safety
	Belt overloaded	Operate belt feed system at design capacity or less

Trouble	Probable causes	Corrective measures
Excessive belt sag between rollers	Take-up tensions too low	Recheck conveyor design and increase take-up mass if possible
	Roller spacing too high	Adjust roller pitch to limit sag to 1%–3% of pitch.
	Erratic loading of belt causing high load points	Improve feed system; regulate feed; maintain clean feed chutes

Inverted V decking consists of metal sheet arranged in inverted V shape between the carrying and return runs of a belt so that the spilled material from the carrying run of the belt is deflected and slide off to the outside of the belt conveyor. Thus, these plates protect the return run of the belt from spilled material.

Bibliography

Belt Conveyors for Bulk Materials, Prepared by the Engineering Conference of the Conveyor Equipment Manufacturers Association, 7th Edition, Naples, Florida, USA, 15 April 2014.

Belt Conveyor idler Installation Instructions & Troubleshooting, 2017, www.ppi-global.com

Belt Conveyor Installation, Operation, Maintenance and Safety manual, Continental Belt Conveyor Systems, Canada, www.continentalconveyor.ca

Belt Conveyor Installation and Operational Maintenance Manual, CEMA Document 2004, Orthman Conveying Systems, Lexington, www.orthman.com

Belt Conveyor Troubleshooting guide, www.ckit.co.za

Conveyor Belt Installation, Maintenance and troubleshooting guide, Continental ContiTech, Fairlawn, USA, www.continental-industry.com

Conveyor Belt Manual, Fenner Dunlop, Conveyor Belting Americas, www.fennerdunlopamericas.com *Conveyor*

Conveyor Handbook, Fenner Dunlop, Conveyor Belting Australia, June 2009, www.fennerdunlop.com

Instruction manual, 2000, Chatland Material Handling Solutions, Humboldt, www.chatland.com

Jurandir Primo, PE: *Belt Conveyors for Bulk Materials Practical Calculations* www.PDHonline.org, 2012.

Ramond.A.Kulwiec, Editor: *Materials Handling Handbook*, American Society of Mechanical Engineers.; International Material Management Society, Wiley & Sons, 1985.

T.S. Kasturi, M.C. Ranga Rajan & T. Krishna Rajan: *Conveyor Components, Operation, Maintenance, Failure Analysis*, PR Business Services, No.49, K.B.Dhasan Road, Alwarpet, Madras, June 1994.

T.S. Kasturi: *Rollers for Material Handling Vol.I*, PR Business Services, No.49, K.B.Dhasan Road, Alwarpet, Madras, October 1990.

T.S. Kasturi: *Conveyor Belt Cleaning Mechanism, Vol.I*, PR Business Services, No.49, K.B.Dhasan Road, Alwarpet, Madras, July 1992.

Z.R.P. Kolacz: *Kevlar Aramid as a Reinforcing Fibre in Conveyor Belts*, Bulk solids handling, Volume 3, Number 4, November 1983, Trans Tech Publications, Clausthal-Zellerfeld, Germany.

H.P. Lachmann: *A Survey on Present-Day Conveyor Belt Technology*, Bulk solids handling, Volume 4, Number 4, December 1984, Trans Tech Publications, Clausthal-Zellerfeld, Germany.

Don McGlinchey: *Bulk Solids Handling: Equipment Selection and Operation*, Blackwell Publishing Ltd, UK, 2008.

Albert Rappen: *Systematic cleaning of Belt Conveyors*, Bulk solids handling, Volume 4, Number 1, March 1984, Trans Tech Publications, Clausthal-Zellerfeld, Germany.

Dietmar Schulze: *Flow Properties of Powders and Bulk Solids*, 2006–2011.

Shah K.P: *Construction and Maintenance of Belt Conveyors for Coal and Bulk Material Handling Plants,* www.practicalmaintenance.net, April 2018.

R. Todd Swinderman, Andrew D. Marti, Larry J. Goldbeck, Daniel Marshall, Mark G. Strebel: *FOUNDATIONS, The Practical Resource for Cleaner, Safer, More Productive Dust & Material Control,* Fourth Edition, Martin Engineering Company, Neponset, Illinois, USA, 2012.

C.R. Woodcock & J.S. Mason: *Bulk Solids Handling, An Introduction to the Practice and Technology,* Leonard Hill Glasgow and London Published in the USA by Chapman and Hall, Newyork, 1987.

Printed in the United States
by Baker & Taylor Publisher Services